Developments in Geotechnical Engineering 8

ROCK DYNAMICS AND GEOPHYSICAL EXPLORATION

Introduction to Stress Waves in Rocks

Further titles in this series:

1. *G. SANGLERAT*
THE PENETROMETER AND SOIL EXPLORATION

2. *Q. ZARUBA AND V. MENCL*
LANDSLIDES AND THEIR CONTROL

3. *E.E. WAHLSTROM*
TUNNELING IN ROCK

4A. *R. SILVESTER*
COASTAL ENGINEERING, I
Generation, Propagation and Influence of Waves

4B. *R. SILVESTER*
COASTAL ENGINEERING, II
Sedimentation, Estuaries, Tides, Effluents, Modelling

5. *R.N. YONG AND B.P. WARKENTIN*
SOIL PROPERTIES AND BEHAVIOUR

6. *E.E. WAHLSTROM*
DAMS, DAM FOUNDATIONS, AND RESERVOIR SITES

7. *W.F. CHEN*
LIMIT ANALYSIS AND SOIL PLASTICITY

Developments in Geotechnical Engineering 8

ROCK DYNAMICS AND GEOPHYSICAL EXPLORATION

Introduction to Stress Waves in Rocks

by

LEIF N. PERSEN

Professor, Institutt for Mekanikk, Universitetet i Trondheim,
Trondheim, Norway

ELSEVIER SCIENTIFIC PUBLISHING COMPANY
Amsterdam Oxford New York 1975

ELSEVIER SCIENTIFIC PUBLISHING COMPANY
335 Jan van Galenstraat
P.O. Box 211, Amsterdam, The Netherlands

AMERICAN ELSEVIER PUBLISHING COMPANY, INC.
52 Vanderbilt Avenue
New York, New York 10017

Library of Congress Card Number: 74-21865

ISBN 0-444-41284-0

Printed in The Netherlands

FOREWORD

For centuries mountains and rock sites were seen merely as almost insurmountable obstacles to development. Even though mines have been known for centuries, it is only a rather recent development to consider rock sites as being something more than a mere inconvenience, and to use them for practical purposes such as underground storage rooms, built-in hydro-power plants, etc. Our ability to use rock sites for practical purposes is due to an ever-increasing technological advance in excavation techniques. With this trend, which it may be assumed will continue to develop in the future, civil engineers and geologists will have an increasing need for a theory which enables them to make predictions in the same way as they are able to do in other fields of engineering activities.

Our present excavation techniques include blasting, which in its turn sets off stress waves which are transmitted through the rock site, and which may strike neighboring installations with destructive force. The study of stress waves in rocks therefore becomes increasingly important, as more and more installations may be found in the neighborhood of the location where blasting occurs. A good knowledge of the way in which such stress waves are transmitted is therefore indispensible for the engineer and the geologist collaborating in such operations.

The present text is an attempt by the author to draw upon his experience in the field to present a theoretical tool for predictions in such cases. The text is built up in such a way that the first seven chapters deal with theory alone and the way in which this theory is deduced, beginning with fundamental principles. Two types of material are mainly considered; the ideally elastic material and the so-called Kelvin material. The last six chapters are concerned with experiments concerning stress waves in rocks, the way in which such experiments may be set up, and the way in which they may be interpreted in the light of theoretical knowledge. It is emphasized that the part dealing with the theoretical aspects of the propagation of stress waves in solids is by no means exhaustive. The wish to make the text more or less self-contained has been the author's guide in this matter, and the reader is referred to the extensive literature in the field for further studies. The text is not exactly elementary, but it is to be considered as an introductory text which may be taught in a one-semester course for graduate students. Only rather simple geometries are considered to bring out fundamental facts. It is also full of numerical examples, which give experimental data in detail, together with their evaluation, so that the reader himself can use these examples to verify that he has a good understanding of the basic ideas.

The author wishes to acknowledge his indebtedness to Dr. Trinks of the German Defence Ministry, for his permission to use the experimental results as examples in this book, and for his enthusiasm and encouragement through numerous discussions during a 10-year period, during which the author served as a scientific advisor for a research project in this field. The author is also very much indebted to A/S Norconsult, Oslo, whose crew performed all the experiments referred to in this text, and who also designed and developed the instrumentation necessary for the successful completion of the field experiments. The author also wishes to thank the National Science Foundation and the University of Rhode Island for making it possible for him, through a grant as a Visiting Senior Scientist, to prepare this text. Without the financial support that I received in this way, the completion of this manuscript would not have been possible. I would also like to thank Professor H. Kolsky, Brown University for his kind suggestions after having read the manuscript. Last, but not least, I am deeply indebted to Professor W. Prager for his encouragement and interest in the work that I have been doing.

Because the present literature on stress waves in rocks is almost non-existent, it is the author's hope that this text will find its way into the engineering curriculum in civil engineering and geology.

University of Rhode Island, May 1973
University of Trondheim, May 1974

LEIF N. PERSEN

CONTENTS

Chapter 1

FUNDAMENTAL EQUATIONS

INTRODUCTORY REMARKS

It is the scope of the present chapter to outline briefly the basic concepts in the analysis of stresses and deformations and then proceed to establish the fundamental equations used to examine the propagation of stress waves in certain types of solids. It is not the intention to give a detailed study of this field. For this purpose the reader is referred to the existing extensive literature.

THE STATE OF STRESS

The concept of a state of stress at a point in a solid under some kind of loading conditions is usually arrived at by initially fixing the attention at a given point in the solid. Choosing an arbitrary reference system (x, y, z) one may then imagine a plane cut through the solid through the point under consideration, and one part of the solid removed. The effect of the removed part on the remaining part may be represented by a force, which is the resultant of a mostly unevenly distributed "stress" (force per unit area) over the area of the cut. At each point of the cut the stress, σ, will have a direction and a magnitude (as a vector) and one may therefore decompose it into one component σ_n (normal stress) normal to the cut, and one component τ (shear stress) in the plane of the cut.

If the cut is supposed to be parallel to the yz-plane, it is now possible to decompose shear stress τ in two components, τ_{xy} and τ_{xz}, so that τ_{xy} has the direction of the y-axis (second subscript) and lies in the same plane. The normal parallel to the x-axis (first subscript). Consequently the second component acts in the direction of the z-axis and lies in the same plane. The normal stress will then be denoted by σ_x, thereby indicating its direction by the subscript. At the point under consideration there are thus three quantities, σ_x, τ_{xy} and τ_{xz}, which characterize the stress there *provided the cut is parallel to the* yz-*plane.*

Now imagine the cut is made along a plane parallel to the zx-plane, the same line of reasoning may be applied, and again under this assumption the stress at the point under consideration is defined by the three quantities, σ_y, τ_{yz} and τ_{yx}. It is then only a matter of routine to conclude that the stress at the point will be σ_z, τ_{zx} and τ_{zy} when the cut is made parallel to the xy-plane.

From the outline above it may be concluded that the stress at the point under consideration is dependent on the orientation of the cut through that point. Analysis shows that, if the nine stresses given above are known at a point, then the stresses on any cut with any orientation may be computed from these nine quantities. Thus, the state of stress at a point is said to be completely defined if the stress tensor, **S**, is known whereby the nine quantities above are said to be the elements of this tensor:

$$\mathbf{S} = \begin{bmatrix} \sigma_x & \tau_{xy} & \tau_{xz} \\ \tau_{yx} & \sigma_y & \tau_{yz} \\ \tau_{zx}, & \tau_{zy} & \sigma_z \end{bmatrix} \tag{1.1}$$

The state of stress given in 1.1 may formally be divided into three parts, i.e.:

$$\mathbf{S} = \mathbf{S}_0 + \mathbf{S}_\mathrm{d} + \mathbf{S}_\mathrm{a} \tag{1.2}$$

where:

$$\mathbf{S}_0 = \begin{bmatrix} \tfrac{1}{3}I_1 & 0 & 0 \\ 0 & \tfrac{1}{3}I_1 & 0 \\ 0 & 0 & \tfrac{1}{3}I_1 \end{bmatrix} \tag{1.3}$$

with:

$$I_1 = \sigma_x + \sigma_y + \sigma_z \tag{1.4}$$

Furthermore:

$$\mathbf{S}_\mathrm{d} = \begin{bmatrix} \sigma_x - \tfrac{1}{3}I_1 & \tfrac{1}{2}(\tau_{xy} + \tau_{yx}) & \tfrac{1}{2}(\tau_{xz} + \tau_{zx}) \\ \tfrac{1}{2}(\tau_{xy} + \tau_{yx}) & \sigma_y - \tfrac{1}{3}I_1 & \tfrac{1}{2}(\tau_{yz} + \tau_{zy}) \\ \tfrac{1}{2}(\tau_{xz} + \tau_{zx}) & \tfrac{1}{2}(\tau_{yz} + \tau_{zy}) & \sigma_z - \tfrac{1}{3}I_1 \end{bmatrix} \tag{1.5}$$

and:

$$\mathbf{S}_\mathrm{a} = \begin{bmatrix} 0 & \tfrac{1}{2}(\tau_{xy} - \tau_{yx}) & \tfrac{1}{2}(\tau_{xz} - \tau_{zx}) \\ \tfrac{1}{2}(\tau_{yx} - \tau_{xy}) & 0 & \tfrac{1}{2}(\tau_{yz} - \tau_{zy}) \\ \tfrac{1}{2}(\tau_{zx} - \tau_{xz}) & \tfrac{1}{2}(\tau_{zy} - \tau_{yz}) & 0 \end{bmatrix} \tag{1.6}$$

The state of stress given by $\mathbf{S}_0$ in 1.3 is generally recognized as a state of "hydrostatic pressure". It is a state in which only normal stresses occur on any plane, and these normal stresses are all of equal magnitude. The state of stress given by $\mathbf{S}_\mathrm{a}$ in 1.6 is (with the exception of Cosserat-media) interpreted as a state of zero stress, because it may be shown that:

$$\tau_{xy} = \tau_{yx}, \qquad \tau_{yz} = \tau_{zy}, \qquad \tau_{zx} = \tau_{xz} \tag{1.7}$$

whereupon the state of stress at a point is finally divided into two parts only:

$$\mathbf{S} = \mathbf{S}_0 + \mathbf{S}_\mathrm{d} \tag{1.8}$$

where now (in view of 1.7):

$$\mathbf{S}_\mathrm{d} = \begin{bmatrix} \sigma_x - \frac{1}{3}I_1 & \tau_{xy} & \tau_{xz} \\ \tau_{xy} & \sigma_y - \frac{1}{3}I_1 & \tau_{yz} \\ \tau_{xz} & \tau_{yz} & \sigma_z - \frac{1}{3}I_1 \end{bmatrix} \tag{1.9}$$

The tensor, $\mathbf{S}_\mathrm{d}$, is generally called "the stress deviator", because it gives the degree to which a given state of stress deviates from the hydrostatic one.

It should be noted that the six stress components are originally conceived of as functions of space, given by the coordinates x, y and z, but they may also be functions of time as will be the case throughout this text.

THE STATE OF DEFORMATION

For the description of the state of deformation of a deformable medium the concept of a displacement vector, $\vec{s}$, is used. Its components u, v and w, in the x-, y- and z-directions respectively, are assumed to be continuous and differentiable functions of the space coordinates and the time t. The deformation is thereby specified as a vector field.

By means of the displacement vector, it is possible to describe the state of deformation in the neighborhood of a point. Locating the origin of the coordinate system at an arbitrary point O, the displacement of which is given by components u, v and w, one may specify the displacement of the neighboring point O_1 in Fig. 1

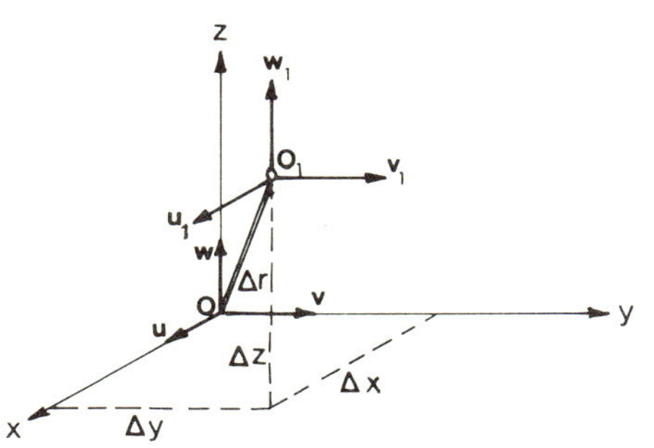

Fig. 1. Relative displacement of two neighboring points.

by u_1, v_1 and w_1 where:

$$u_1 = u + \frac{\partial u}{\partial x}\Delta x + \frac{\partial u}{\partial y}\Delta y + \frac{\partial u}{\partial z}\Delta z + \text{higher-order terms}$$

$$v_1 = v + \frac{\partial v}{\partial x}\Delta x + \frac{\partial v}{\partial y}\Delta y + \frac{\partial v}{\partial z}\Delta z + \text{higher-order terms}$$

$$w_1 = w + \frac{\partial w}{\partial x}\Delta x + \frac{\partial w}{\partial y}\Delta y + \frac{\partial w}{\partial z}\Delta z + \text{higher-order terms}$$

If the higher-order terms are neglected in these expressions, the mutual displacement of the two points will be:

$$\left.\begin{aligned} u_1 - u &= \frac{\partial u}{\partial x}\Delta x + \frac{\partial u}{\partial y}\Delta y + \frac{\partial u}{\partial z}\Delta z \\ v_1 - v &= \frac{\partial v}{\partial x}\Delta x + \frac{\partial v}{\partial y}\Delta y + \frac{\partial v}{\partial z}\Delta z \\ w_1 - w &= \frac{\partial w}{\partial x}\Delta x + \frac{\partial w}{\partial y}\Delta y + \frac{\partial w}{\partial z}\Delta z \end{aligned}\right| \qquad [1.10]$$

From these expressions it is clear that the state of deformation around any point is completely determined by the nine coefficients listed in the following matrix, the elements of which form the elements of the deformation tensor **D**:

$$\mathbf{D} = \begin{bmatrix} \frac{\partial u}{\partial x} & \frac{\partial u}{\partial y} & \frac{\partial u}{\partial z} \\ \frac{\partial v}{\partial x} & \frac{\partial v}{\partial y} & \frac{\partial v}{\partial z} \\ \frac{\partial w}{\partial x} & \frac{\partial w}{\partial y} & \frac{\partial w}{\partial z} \end{bmatrix} \qquad [1.11]$$

This tensor can be divided into three parts in the same way as the stress tensor:

$$\mathbf{D} = \mathbf{D}_0 + \mathbf{D}_\mathrm{d} + \mathbf{D}_\mathrm{a} \qquad [1.12]$$

Here:

$$\mathbf{D}_0 = \tfrac{1}{3}\begin{bmatrix} e & 0 & 0 \\ 0 & e & 0 \\ 0 & 0 & e \end{bmatrix} \qquad [1.13]$$

$$\mathbf{D}_\mathrm{d} = \begin{bmatrix} \frac{\partial u}{\partial x} - \frac{1}{3}e & \frac{1}{2}\left(\frac{\partial u}{\partial y} + \frac{\partial v}{\partial x}\right) & \frac{1}{2}\left(\frac{\partial u}{\partial z} + \frac{\partial w}{\partial x}\right) \\ \frac{1}{2}\left(\frac{\partial v}{\partial x} + \frac{\partial u}{\partial y}\right) & \frac{\partial v}{\partial y} - \frac{1}{3}e & \frac{1}{2}\left(\frac{\partial v}{\partial z} + \frac{\partial w}{\partial y}\right) \\ \frac{1}{2}\left(\frac{\partial w}{\partial x} + \frac{\partial u}{\partial z}\right) & \frac{1}{2}\left(\frac{\partial w}{\partial y} + \frac{\partial v}{\partial z}\right) & \frac{\partial w}{\partial z} - \frac{1}{3}e \end{bmatrix} \qquad [1.14]$$

$$\mathbf{D}_\mathrm{a} = \begin{bmatrix} 0 & \frac{1}{2}\left(\frac{\partial u}{\partial y} - \frac{\partial v}{\partial x}\right) & \frac{1}{2}\left(\frac{\partial u}{\partial z} - \frac{\partial w}{\partial x}\right) \\ \frac{1}{2}\left(\frac{\partial v}{\partial x} - \frac{\partial u}{\partial y}\right) & 0 & \frac{1}{2}\left(\frac{\partial v}{\partial z} - \frac{\partial w}{\partial y}\right) \\ \frac{1}{2}\left(\frac{\partial w}{\partial x} - \frac{\partial u}{\partial z}\right) & \frac{1}{2}\left(\frac{\partial w}{\partial y} - \frac{\partial v}{\partial z}\right) & 0 \end{bmatrix} \qquad [1.15]$$

where e is the first invariant of the deformation tensor:

$$e = \frac{\partial u}{\partial x} + \frac{\partial v}{\partial y} + \frac{\partial w}{\partial z} \qquad [1.16]$$

The state of deformation may be described by specifying the three components of $\mathbf{D}_\mathrm{a}$, five of the components of $\mathbf{D}_\mathrm{d}$ (one of the elements on the main diagonal being given implicitly when two of them are known) and the one element in $\mathbf{D}_0$, which again means nine components in all. The reason for this division is that a physical interpretation of the different parts may easily be given.

By means of 1.10 it is possible to deduce an expression for the relative extension, ϵ, of the distance, $\Delta\vec{r}$, between O and O_1 in Fig. 1.

$$(1 + \epsilon)^2 (\Delta r)^2 = (\Delta x + u_1 - u)^2 + (\Delta y + v_1 - v)^2 + (\Delta z + w_1 - w)^2 \qquad [1.17]$$

By neglecting higher-order terms this expression may be reduced to:

$$\epsilon = \frac{\partial u}{\partial x} l^2 + \frac{\partial v}{\partial y} m^2 + \frac{\partial w}{\partial z} n^2 + \left(\frac{\partial v}{\partial x} + \frac{\partial u}{\partial y}\right) lm + \left(\frac{\partial w}{\partial y} + \frac{\partial v}{\partial z}\right) mn + \left(\frac{\partial u}{\partial z} + \frac{\partial w}{\partial x}\right) nl \qquad [1.18]$$

where l, m and n represent the directional cosines of vector $\Delta\vec{r}$:

$$l = \frac{\Delta x}{\Delta r}, \quad m = \frac{\Delta y}{\Delta r}, \quad n = \frac{\Delta z}{\Delta r} \qquad [1.19]$$

Fig. 2. shows an element with sides a, b, and c undergoing a change of volume. The extension of the edges are Δa, Δb and Δc and with reference to eq. 1.18 the relative extensions will be:

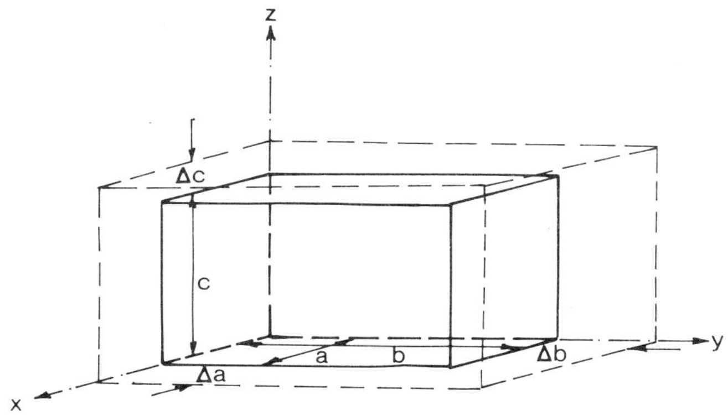

Fig. 2. Volume change of an element.

$$\begin{aligned} \epsilon_x &= \frac{\Delta a}{a} = \frac{\partial u}{\partial x} \\ \epsilon_y &= \frac{\Delta b}{b} = \frac{\partial v}{\partial y} \\ \epsilon_z &= \frac{\Delta c}{c} = \frac{\partial w}{\partial z} \end{aligned} \qquad [1.20]$$

The relative change in volume will be:

$$\frac{\Delta V}{V} = \frac{a(1+\epsilon_x)\,b(1+\epsilon_y)\,c(1+\epsilon_z) - abc}{abc} =$$

$$= \frac{\partial u}{\partial x} + \frac{\partial v}{\partial y} + \frac{\partial w}{\partial z} + \text{higher-order terms} \qquad [1.21]$$

Neglecting higher-order terms one may conclude that e in eq. 1.16 represents the relative change of volume. It is further observed that this change of volume is represented by the sum of the elements on the main diagonal of the deformation tensor, **D**. This is the same as the sum of the corresponding elements of $\mathbf{D}_0$, whereas the sum of the corresponding elements of $\mathbf{D}_d$ and $\mathbf{D}_a$ is zero. Thus the two latter parts of the total deformation take place without change of volume. Because $\mathbf{D}_0$ exhibits spherical symmetry, it is clear that it represents a change of volume in such a way that the shape of an element is geometrically similar before and after deformation. This means that $\mathbf{D}_0$ represents that part of the total deformation which may be described as "a change of volume at constant shape" or "*a shape-true volume change.*"

In order to find a physical interpretation of $\mathbf{D}_a$, a rigid body rotating an angle, $\vec{\Omega}$, about a point is considered as shown in Fig. 3. If an arbitrary point in the body

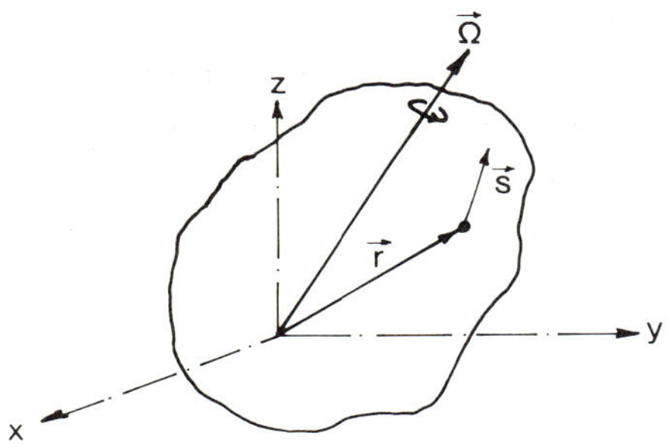

Fig. 3. Rigid body rotation.

is given by its position vector $\vec{r}$, the displacement, $\vec{s}$, of this point will be given as:

$$\vec{s} = \vec{\Omega} \times \vec{r} \qquad [1.22]$$

Denoting the components of $\vec{\Omega}$ and $\vec{r}$ by $\omega_x, \omega_y, \omega_z$ and x, y, z respectively, 1.22 will upon being resolved into components give:

$$\left.\begin{aligned} u &= \omega_y z - \omega_z y \\ v &= \omega_z x - \omega_x z \\ w &= \omega_x y - \omega_y x \end{aligned}\right| \qquad [1.23]$$

whereupon:

$$\left.\begin{aligned} &\frac{\partial u}{\partial x} = 0, && \frac{\partial v}{\partial x} = +\omega_z, && \frac{\partial w}{\partial x} = -\omega_y \\ &\frac{\partial u}{\partial y} = -\omega_z, && \frac{\partial v}{\partial y} = 0, && \frac{\partial w}{\partial y} = +\omega_x \\ &\frac{\partial u}{\partial z} = +\omega_y, && \frac{\partial v}{\partial z} = -\omega_x, && \frac{\partial w}{\partial z} = 0 \end{aligned}\right| \qquad [1.24]$$

That part of the total deformation which is represented by $\mathbf{D}_a$ will in the present case be:

$$\mathbf{D}_a = \begin{bmatrix} 0 & -\omega_z & +\omega_y \\ +\omega_z & 0 & -\omega_x \\ -\omega_y & +\omega_x & 0 \end{bmatrix} \qquad [1.25]$$

whereas $\mathbf{D}_0$ and $\mathbf{D}_d$ will be, identically, zero. This means that $\mathbf{D}_a$ represents that part of the total deformation which corresponds to a rotation as a rigid body. This leaves $\mathbf{D}_d$ to be interpreted as that part of the total deformation which consists of a change of shape at constant volume, or *"a volume-true shape change"* (distortion). The division of the deformation tensor **D** into three parts [1.12] has thus served the purpose of dividing an arbitrary deformation around a point into: (1) a shape-true change of volume, (2) a volume-true change of shape, and (3) a rotation as a rigid body.

EQUATIONS OF MOTION

The equations governing the propagation of stress waves in a solid medium are now established by imagining an element of the solid at a fixed time isolated from its surroundings, and replacing the influence of these surroundings on the element by the proper stresses. The element is supposed to be so small that to the first order of accuracy the stresses may be thought of as evenly distributed over the surfaces. The element is shown in Fig. 4 and it is now rather easy to show that the surroundings will excert a force on the element, the components of which will be:

$$\begin{aligned} &\text{in the } x\text{-direction: } \Delta x\ \Delta y\ \Delta z \left(\frac{\partial \sigma_x}{\partial x} + \frac{\partial \tau_{xy}}{\partial y} + \frac{\partial \tau_{xz}}{\partial z}\right) \\ &\text{in the } y\text{-direction: } \Delta x\ \Delta y\ \Delta z \left(\frac{\partial \tau_{xy}}{\partial x} + \frac{\partial \sigma_y}{\partial y} + \frac{\partial \tau_{yz}}{\partial z}\right) \\ &\text{in the } z\text{-direction: } \Delta x\ \Delta y\ \Delta z \left(\frac{\partial \tau_{xz}}{\partial x} + \frac{\partial \tau_{yz}}{\partial y} + \frac{\partial \sigma_z}{\partial z}\right) \end{aligned} \qquad [1.26]$$

The element, which is moving, but which is considered at an arbitrary instant, will undergo accelerations. The d'Alemberts inertia force may be expressed as:

$$-\rho\Delta \cdot \Delta y\ \Delta z \frac{\partial^2 \vec{s}}{\partial t^2} \qquad [1.27]$$

and the equations of dynamic equilibrium will then be:

$$\begin{aligned} \frac{\partial \sigma_x}{\partial x} + \frac{\partial \tau_{xy}}{\partial y} + \frac{\partial \tau_{xz}}{\partial z} &= \rho \frac{\partial^2 u}{\partial t^2} \\ \frac{\partial \tau_{xy}}{\partial x} + \frac{\partial \sigma_y}{\partial y} + \frac{\partial \tau_{yz}}{\partial z} &= \rho \frac{\partial^2 v}{\partial t^2} \\ \frac{\partial \tau_{xz}}{\partial x} + \frac{\partial \tau_{yz}}{\partial y} + \frac{\partial \sigma_z}{\partial z} &= \rho \frac{\partial^2 w}{\partial t^2} \end{aligned} \qquad [1.28]$$

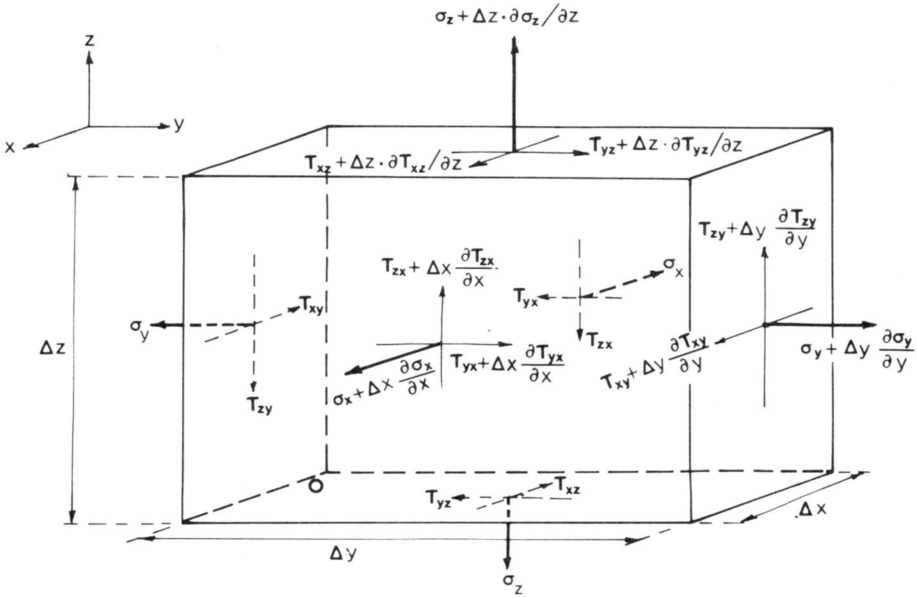

Fig. 4. Stresses acting on an element.

These equations of motions cannot be developed any further before the state of stress somehow is related to the state of deformation, whereby the stresses will be expressed in terms of the components u, v and w of the displacement vector $\vec{s}$.

THE PHENOMENOLOGICAL RELATION (ELASTIC MATERIAL)

The relation connecting the state of stress to the state of deformation is frequently called the phenomenological relation. It is often thought of as being a fundamental equation in the thermodynamic sense, and the coefficients appearing in the relation are regarded as parameters characterizing the properties of the material. When dealing with an elastic material, this point of view may to some extent be acceptable, but as will be shown at a much later stage, the phenomenological relation may rather be conceived of as a relation characteristic of the "process" (in the thermodynamical sense).

The phenomenological relation is in its nature a relation based on experimental evidence, and as such characterizes the process whereby material properties of the medium also enter the relation. It may, however, be introduced as a hypothesis, whereby an ideal case is established. It is then a matter of experimental verification to find out to what extent a theory based on such a hypothesis represents physical reality.

Usually the phenomenological relations used are linear ones. The case of an ideally elastic material is one such case, where in terms of the stress tensor and the deformation tensor the relation is stated as:

$$\left.\begin{aligned} \mathbf{S}_0 &= k_0\,\mathbf{D}_0 \\ \mathbf{S}_\mathrm{d} &= k_\mathrm{d}\,\mathbf{D}_\mathrm{d} \\ \mathbf{S}_\mathrm{a} &= k_\mathrm{a}\,\mathbf{D}_\mathrm{a} \end{aligned}\right| \qquad [1.29]$$

It is of course observed, that $\mathbf{S}_\mathrm{a} = 0$, whereas in general $\mathbf{D}_\mathrm{a} \neq 0$ which leads to $k_\mathrm{a} = 0$. Thus, for the ideally elastic case, only the two constants k_0 and k_d will enter the relation. These constants are:

$$k_0 = 2G + 3\lambda, \quad k_\mathrm{d} = 2G \qquad [1.30]$$

where G and λ are the Lamé constants. Expressed in terms of the perhaps more familiar constants, the constant λ may be written:

$$\lambda \equiv \frac{\nu E}{(1+\nu)(1-2\nu)} \equiv \frac{2\nu G}{1-2\nu} \qquad [1.31]$$

where E = Young's modulus, ν = Poisson's ratio, and G = the shear modulus.

Eqs. 1.29 may now in view of 1.30 and 1.31 be explicitly expressed as follows:

$$I_1 = \sigma_x + \sigma_y + \sigma_z = (2G + 3\lambda)\left(\frac{\partial u}{\partial x} + \frac{\partial v}{\partial y} + \frac{\partial w}{\partial z}\right) \qquad [1.32]$$

$$\left.\begin{aligned} \sigma_x &= 2G\frac{\partial u}{\partial x} + \lambda\left(\frac{\partial u}{\partial x} + \frac{\partial v}{\partial y} + \frac{\partial w}{\partial z}\right) \\ \sigma_y &= 2G\frac{\partial v}{\partial y} + \lambda\left(\frac{\partial u}{\partial x} + \frac{\partial v}{\partial y} + \frac{\partial w}{\partial z}\right) \\ \sigma_z &= 2G\frac{\partial w}{\partial z} + \lambda\left(\frac{\partial u}{\partial x} + \frac{\partial v}{\partial y} + \frac{\partial w}{\partial z}\right) \end{aligned}\right| \qquad [1.33]$$

$$\left.\begin{aligned} \tau_{xy} &= G\left(\frac{\partial u}{\partial y} + \frac{\partial v}{\partial x}\right) \\ \tau_{yz} &= G\left(\frac{\partial v}{\partial z} + \frac{\partial w}{\partial y}\right) \\ \tau_{zx} &= G\left(\frac{\partial w}{\partial x} + \frac{\partial u}{\partial z}\right) \end{aligned}\right| \qquad [1.34]$$

One will observe that the information needed to develop 1.28 further, is given in 1.33 and 1.34, and it is only a matter of routine algebra to establish the fundamental equations describing the wave propagation in an elastic material.

THE PHENOMENOLOGICAL RELATION (KELVIN MATERIAL)

In discussing the phenomenological relation for an ideally elastic material, it was pointed out that this relation could be conceived of as a relation describing the process that takes place. One particular feature of this process in the elastic case is the absence of energy dissipation. The phenomenological relation for the process in an elastic material has built into it the assumption that the stresses are independent of the rate of deformation. It will be shown that this situation leads to unacceptable conclusions when applied to wave propagation in rocks (and other materials for that matter).

The simplest possible assumption which will remedy this situation is to assume a time dependence in the phenomenological relation. This may be accomplished as follows:

$$\left.\begin{aligned} \mathbf{S}_0 &= k_0 \left(1 + \eta_0 \frac{\partial}{\partial t}\right) \mathbf{D}_0 \\ \mathbf{S}_d &= k_d \left(1 + \eta_d \frac{\partial}{\partial t}\right) \mathbf{D}_d \end{aligned}\right| \qquad [1.35]$$

Such a phenomenological relation states that the state of stress does not only depend on the state of deformation but also on the rate at which the latter changes.

One may pause here for a moment to contemplate the rock as a material. It is obvious that a body of rock as it appears in nature is not a homogeneous material. It exhibits very often a wavy stratification, it is cut by cracks and it may even exhibit geological characteristics which change with location. When stress waves pass through a material like this, an apparent energy dissipation is observed. If, however, a small sample of the rock is brought to the laboratory for testing, this sample may exhibit an almost elastic behavior.

With this in mind, an attempt to solve the problem might be conceived of whereby a study of the transmission of stress waves through stratified material, through inhomogeneous materials or through cracks would furnish a starting point. One might then proceed to assume some sort of statistical distribution in the rock of these characteristics of the rock, and compute their overall effect on the propagation of stress waves. In this way one would be in an analogous position as when e.g. the viscosity of a fluid is computed from some microscopical model of the fluid. The success of the exercise can only be judged from a confrontation of the result with the macroscopically measured viscosity. The logical and easiest conclusion which may be drawn is therefore, in the case considered here, to assume a "viscous" damping effect (expressed here by η_0 and η_d) in the first place and not describe in detail how this damping is brought about.

Now, if the phenomenological relation [1.35] is examined in detail, the conclusion which seems to be obvious is that it assumes the rock to be a homogeneous visco-elastic material of the Kelvin type in its simplest form. However, relation 1.35 is to be conceived of as a relation describing the process, and thus the constants appearing in the relation (k_0, k_d, η_0 and η_d) are no longer to be interpreted as parameters characterizing the material only. These preliminary remarks will be reconsidered when experimental results may be drawn upon.

Returning to 1.35, the equations are expressed explicitly as follows:

$$I_1 = \sigma_x + \sigma_y + \sigma_z = k_0\left(1 + \eta_0\frac{\partial}{\partial t}\right)\left(\frac{\partial u}{\partial x} + \frac{\partial v}{\partial y} + \frac{\partial w}{\partial z}\right) \qquad [1.36]$$

$$\left.\begin{aligned}
\sigma_x &= k_d\left(1 + \eta_d\frac{\partial}{\partial t}\right)\frac{\partial u}{\partial x} + \frac{1}{3}\left[(k_0 - k_d) + (k_0\eta_0 - k_d\eta_d)\frac{\partial}{\partial t}\right]\operatorname{div}\vec{s}\\
\sigma_y &= k_d\left(1 + \eta_d\frac{\partial}{\partial t}\right)\frac{\partial v}{\partial y} + \frac{1}{3}\left[(k_0 - k_d) + (k_0\eta_0 - k_d\eta_d)\frac{\partial}{\partial t}\right]\operatorname{div}\vec{s}\\
\sigma_z &= k_d\left(1 + \eta_d\frac{\partial}{\partial t}\right)\frac{\partial w}{\partial z} + \frac{1}{3}\left[(k_0 - k_d) + (k_0\eta_0 - k_d\eta_d)\frac{\partial}{\partial t}\right]\operatorname{div}\vec{s}
\end{aligned}\right\} \qquad [1.37]$$

$$\left.\begin{aligned}
\tau_{xy} &= \tfrac{1}{2}k_d\left(1 + \eta_d\frac{\partial}{\partial t}\right)\left(\frac{\partial u}{\partial y} + \frac{\partial v}{\partial x}\right)\\
\tau_{yz} &= \tfrac{1}{2}k_d\left(1 + \eta_d\frac{\partial}{\partial t}\right)\left(\frac{\partial v}{\partial z} + \frac{\partial w}{\partial y}\right)\\
\tau_{zx} &= \tfrac{1}{2}k_d\left(1 + \eta_d\frac{\partial}{\partial t}\right)\left(\frac{\partial w}{\partial x} + \frac{\partial u}{\partial z}\right)
\end{aligned}\right\} \qquad [1.38]$$

Introduction of 1.37 and 1.38 into 1.28 will give the basic equations of motion in their final formulation.

COORDINATE SYSTEMS

The deduction of the basic equations and the introduction of basic concepts have so far been described in Cartesian coordinate systems only. Very often applications involve geometries for which other coordinate systems are more suitable. For a general survey of the basic equations and concepts the reader is referred to the extensive literature on this subject. For the sake of reference some of these are listed below:

(a) Cartesian coordinates (x, y, z)

(1) Deformation tensor:

$$\mathbf{D} = \begin{bmatrix} \epsilon_x & \frac{1}{2}\gamma_{xy} & \frac{1}{2}\gamma_{xz} \\ \frac{1}{2}\gamma_{xy} & \epsilon_y & \frac{1}{2}\gamma_{yz} \\ \frac{1}{2}\gamma_{xz} & \frac{1}{2}\gamma_{yz} & \epsilon_z \end{bmatrix} \qquad [1.39]$$

(2) Displacement vector:

$$\vec{s} = \vec{i}_x u + \vec{i}_y v + \vec{i}_z w \qquad [1.40]$$

(3) Strain:

$$\epsilon_x = \frac{\partial u}{\partial x}, \quad \epsilon_y = \frac{\partial v}{\partial y}, \quad \epsilon_z = \frac{\partial w}{\partial z} \qquad [1.41]$$

(4) Shear strain:

$$\gamma_{xy} = \frac{\partial u}{\partial y} + \frac{\partial v}{\partial x}, \quad \gamma_{yz} = \frac{\partial v}{\partial z} + \frac{\partial w}{\partial y}, \quad \gamma_{zx} = \frac{\partial w}{\partial x} + \frac{\partial u}{\partial z} \qquad [1.42]$$

(5) Equations of motion:

$$\begin{aligned} \frac{\partial \sigma_x}{\partial x} + \frac{\partial \tau_{xy}}{\partial y} + \frac{\partial \tau_{xz}}{\partial z} &= \rho \frac{\partial^2 u}{\partial t^2} \\ \frac{\partial \tau_{xy}}{\partial x} + \frac{\partial \sigma_y}{\partial y} + \frac{\partial \tau_{yz}}{\partial z} &= \rho \frac{\partial^2 v}{\partial t^2} \\ \frac{\partial \tau_{xz}}{\partial x} + \frac{\partial \tau_{yz}}{\partial y} + \frac{\partial \sigma_z}{\partial z} &= \rho \frac{\partial^2 w}{\partial t^2} \end{aligned} \qquad [1.43]$$

(b) Cylindrical coordinates (r, φ, z)

(1) Deformation tensor:

$$\mathbf{D} = \begin{bmatrix} \epsilon_r & \frac{1}{2}\gamma_{r\varphi} & \frac{1}{2}\gamma_{rz} \\ \frac{1}{2}\gamma_{r\varphi} & \epsilon_\varphi & \frac{1}{2}\gamma_{\varphi z} \\ \frac{1}{2}\gamma_{zr} & \frac{1}{2}\gamma_{z\varphi} & \epsilon_z \end{bmatrix} \qquad [1.44]$$

(2) Displacement vector:

$$\vec{s} = \vec{i}_r v_r + \vec{i}_\varphi v_\varphi + \vec{i}_z v_z \qquad [1.45]$$

(3) Strain:

$$\epsilon_r = \frac{\partial v_r}{\partial r}, \quad \epsilon_\varphi = \frac{1}{r}\frac{\partial v_\varphi}{\partial \varphi} + \frac{v_r}{r}, \quad \epsilon_z = \frac{\partial v_z}{\partial z} \qquad [1.46]$$

(4) Shear strain:

$$\left.\begin{aligned} \gamma_{r\varphi} &= \frac{1}{r}\frac{\partial v_r}{\partial \varphi} + r\frac{\partial}{\partial r}\left(\frac{v_\varphi}{r}\right) \\ \gamma_{\varphi z} &= \frac{\partial v_\varphi}{\partial z} + \frac{1}{r}\frac{\partial v_z}{\partial \varphi} \\ \gamma_{zr} &= \frac{\partial v_r}{\partial z} + \frac{\partial v_z}{\partial r} \end{aligned}\right| \qquad [1.47]$$

(5) Equations of motion:

$$\left.\begin{aligned} &\frac{\partial \sigma_r}{\partial r} + \frac{1}{r}\frac{\partial \tau_{r\varphi}}{\partial \varphi} + \frac{\partial \tau_{rz}}{\partial z} + \frac{\sigma_r - \sigma_\varphi}{r} = \rho\frac{\partial^2 v_r}{\partial t^2} \\ &\frac{\partial \tau_{r\varphi}}{\partial r} + \frac{1}{r}\frac{\partial \sigma_\varphi}{\partial \varphi} + \frac{\partial \tau_{\varphi z}}{\partial z} + 2\frac{\tau_{r\varphi}}{r} = \rho\frac{\partial^2 v_\varphi}{\partial t^2} \\ &\frac{\partial \tau_{r\varphi}}{\partial r} + \frac{1}{r}\frac{\partial \tau_{\varphi z}}{\partial \varphi} + \frac{\partial \sigma_z}{\partial z} + \frac{\tau_{rz}}{r} = \rho\frac{\partial^2 v_z}{\partial t^2} \end{aligned}\right| \qquad [1.48]$$

(*c*) *Spherical coordinates* (r, φ, θ)

(1) Deformation tensor:

$$\mathbf{D} = \begin{bmatrix} \epsilon_r & \frac{1}{2}\gamma_{r\varphi} & \frac{1}{2}\gamma_{r\theta} \\ \frac{1}{2}\gamma_{r\varphi} & \epsilon_\varphi & \frac{1}{2}\gamma_{\varphi\theta} \\ \frac{1}{2}\gamma_{r\theta} & \frac{1}{2}\gamma_{\varphi\theta} & \epsilon_\theta \end{bmatrix} \qquad [1.49]$$

(2) Displacement vector:

$$\vec{s} = \vec{i}_r v_r + \vec{i}_\varphi v_\varphi + \vec{i}_\theta v_\theta \qquad [1.50]$$

(3) Strain:

$$\begin{aligned}
\epsilon_r &= \frac{\partial v_r}{\partial r} \\
\epsilon_\varphi &= \frac{1}{r\sin\theta}\frac{\partial v_\varphi}{\partial \varphi} + \frac{v_r}{r} + \frac{v_\theta}{r}\cot\theta \\
\epsilon_\theta &= \frac{1}{r}\frac{\partial v_\theta}{\partial \theta} + \frac{v_r}{r}
\end{aligned} \qquad [1.51]$$

(4) Shear strain:

$$\begin{aligned}
\gamma_{r\varphi} &= \frac{1}{r\sin\theta}\frac{\partial v_r}{\partial \varphi} + r\frac{\partial}{\partial r}\left(\frac{v_\varphi}{r}\right) \\
\gamma_{\varphi\theta} &= \frac{1}{r}\frac{\partial v_\varphi}{\partial \theta} - \frac{v_\varphi}{r}\cot\theta + \frac{1}{r\sin\theta}\frac{\partial v_\theta}{\partial \varphi} \\
\gamma_{\theta r} &= \frac{1}{r}\frac{\partial v_r}{\partial \theta} + r\frac{\partial}{\partial r}\left(\frac{v_\theta}{r}\right)
\end{aligned} \qquad [1.52]$$

(5) Equations of motion:

$$\begin{aligned}
&\frac{\partial \sigma_r}{\partial r} + \frac{1}{r\sin\theta}\frac{\partial \tau_{r\varphi}}{\partial \varphi} + \frac{1}{r}\frac{\partial \tau_{r\theta}}{\partial \theta} + \frac{2\sigma_r - \sigma_\varphi - \sigma_\theta + \tau_{r\theta}\cot\theta}{r} = \rho\frac{\partial^2 v_r}{\partial t^2} \\
&\frac{\partial \tau_{r\varphi}}{\partial r} + \frac{1}{r\sin\theta}\frac{\partial \sigma_\varphi}{\partial \varphi} + \frac{1}{r}\frac{\partial \tau_{\varphi\theta}}{\partial \theta} + \frac{3\tau_{r\varphi} + 2\tau_{\theta\varphi}\cot\theta}{r} = \rho\frac{\partial^2 v_\varphi}{\partial t^2} \\
&\frac{\partial \tau_{r\theta}}{\partial r} + \frac{1}{r\sin\theta}\frac{\partial \tau_{\theta\varphi}}{\partial \varphi} + \frac{1}{r}\frac{\partial \sigma_\theta}{\partial \theta} + \frac{3\tau_{r\theta} + (\sigma_\theta - \sigma_\varphi)\cot\theta}{r} = \rho\frac{\partial^2 v_\theta}{\partial t^2}
\end{aligned} \qquad [1.53]$$

Chapter 2

SOME CHARACTERISTIC FEATURES OF PLANE WAVES (ELASTIC MEDIUM)

INTRODUCTION

It is not within the scope of this chapter to give a broad and comprehensive treatment of the wave propagation in an elastic medium. The intention is rather by selecting specific problems to exhibit through their solution some general aspects of the wave propagation. Because the general aim is to apply the theory to the wave propagation in rocks, the emphasis will be on problems which have some relevance in this respect.

BASIC EQUATIONS IN CARTESIAN COORDINATES

By introducing 1.33 and 1.34 into 1.28 one obtains the following equations:

$$\left.\begin{aligned} G\nabla^2 u + (G+\lambda)\frac{\partial}{\partial x}\left(\frac{\partial u}{\partial x}+\frac{\partial v}{\partial y}+\frac{\partial w}{\partial z}\right) &= \rho\frac{\partial^2 u}{\partial t^2} \\ G\nabla^2 v + (G+\lambda)\frac{\partial}{\partial y}\left(\frac{\partial u}{\partial x}+\frac{\partial v}{\partial y}+\frac{\partial w}{\partial z}\right) &= \rho\frac{\partial^2 v}{\partial t^2} \\ G\nabla^2 w + (G+\lambda)\frac{\partial}{\partial z}\left(\frac{\partial u}{\partial x}+\frac{\partial v}{\partial y}+\frac{\partial w}{\partial z}\right) &= \rho\frac{\partial^2 w}{\partial t^2} \end{aligned}\right| \quad [2.1]$$

These are the basic equations governing the propagation of stress waves in an ideally elastic material. They will serve as the starting point for exhibiting some rather general features of the stress wave propagation in this particular medium.

First it should be observed that the basic equations may be represented in vector form as follows:

$$(2G+\lambda)\,\mathrm{grad}\,(\mathrm{div}\,\vec{s}) - G\,\mathrm{curl}\,(\mathrm{curl}\,\vec{s}) = \rho\frac{\partial^2\vec{s}}{\partial t^2} \quad [2.2]$$

This equation has the advantage of being true in all orthogonal curvilinear coordinates, and conclusions drawn from it will thus have a general character. Two cases in which 2.2 is greatly simplified, may now be readily recognized.

Case 1 (curl $\vec{s} \equiv 0$)

The first case may be introduced by assuming that the state of deformation transmitted by the wave is *irrotational,* i.e. in view of eq. 1.15:

$$\mathbf{D}_a \equiv 0 \qquad [2.3]$$

This is a kinematic restriction placed on the deformation field, and may be expressed as:

$$\text{curl}\,\vec{s} = 0 \qquad [2.4]$$

This condition is identically satisfied if $\vec{s}$ is derived from a scalar potential function $\Phi(x, y, z)$:

$$\vec{s} = \text{grad}\,\Phi \qquad [2.5]$$

It is then easily verified that:

$$\text{div}\,\vec{s} = \text{div}\,(\text{grad}\,\Phi) = \nabla^2\Phi \qquad [2.6]$$

where:

$$\nabla^2 \equiv \frac{\partial^2}{\partial x^2} + \frac{\partial^2}{\partial y^2} + \frac{\partial^2}{\partial z^2} \qquad [2.7]$$

Substitution of 2.4, 2.5 and 2.6 into 2.2 reduces the latter equation to:

$$\nabla^2\Phi = \frac{1}{c^2}\frac{\partial^2\Phi}{\partial t^2} \qquad [2.8]$$

where:

$$c^2 = \frac{2G + \lambda}{\rho} \qquad [2.9]$$

The constant c has the dimension of a velocity, and is usually called the velocity of a compression wave or the velocity of a longitudinal wave. Φ is a scalar potential function whose propagation is described by 2.8.

Case 2 (div $\vec{s} = 0$)

The second case may be introduced by assuming that the state of deformation transmitted by the wave is *equivoluminal,* i.e. in view of 1.21:

$$\text{div}\, \vec{s} = 0 \qquad [2.10]$$

This is again a kinematic restriction placed on the deformation field and it will always be satisfied if $\vec{s}$ is derived from a vector potential function, $\vec{\Psi}$:

$$\vec{s} = \text{curl}\, \vec{\Psi} \qquad [2.11]$$

If now 2.10 and 2.11 are introduced into 2.2, one obtains:

$$-\,\text{curl}\,[\text{curl}\,(\text{curl}\,\vec{\Psi})] = \frac{1}{c_t^2}\frac{\partial^2}{\partial t^2}(\text{curl}\,\vec{\Psi})$$

or:

$$-\,\text{curl}\,(\text{curl}\,\vec{\Psi}) = \frac{1}{c_t^2}\frac{\partial^2 \vec{\Psi}}{\partial t^2} \qquad [2.12]$$

where:

$$c_t^2 = \frac{G}{\rho} \qquad [2.13]$$

The constant c_t is called the shear velocity or the velocity of the tangential wave. $\vec{\Psi}$ is the vector potential function, whose propagation is described by 2.12.

The general case

Having now described the two special cases, it is rather easy to contemplate the general case. Because the equations are linear, a superposition of solutions is always possible, and a uniqueness theorem for the solutions exists. Thus, in general, the displacement vector $\vec{s}$ may be represented by:

$$\vec{s} = \text{grad}\, \Phi + \text{curl}\, \vec{\Psi} \qquad [2.14]$$

and again the basic equations may be divided into two sets, each set of equations describing the propagation of the compression wave and the shear wave. Because always $c > c_t$, it means that, in general, the situation at a given point in the material is such that first the arrival of the compression wave will be noticed and then, somewhat later, the arrival of the shear wave will be felt. The general statements in this paragraph will be further exhibited in a survey of very simple examples.

PLANE WAVES IN THE HALFSPACE

So far the approach to the problem has been more or less theoretical in the way that only the differential equations have been considered. Now, the boundary and initial conditions must be drawn into consideration. Thus, the simplest possible case of a plane wave is considered. Assume an ideally elastic halfspace which exhibits a free surface at the plane $x = 0$. The halfspace is supposed initially to be free of all stresses. At time $t = 0$, it is assumed that an evenly distributed stress, σ_0, is instantaneously applied at the free surface of the halfspace, and is then kept constant for all $t > 0$. The applied stress at the surface is in other words a Heaviside function of time. The geometry of the problem is such that components v and w of the displacement vector $\vec{s}$ are equal to zero and remain so during the process.

Under these conditions component u of the displacement vector $\vec{s}$ will be the only unknown. The basic equations [2.1] will reduce to:

$$\left.\begin{aligned} &\frac{\partial^2 u}{\partial x^2} = \frac{1}{c^2}\frac{\partial^2 u}{\partial t^2} \\ &\frac{\partial u}{\partial y} = \frac{\partial u}{\partial z} = 0 \end{aligned}\right| \qquad [2.15]$$

In this case:

$$\operatorname{div} \vec{s} \neq 0, \quad \operatorname{curl} \vec{s} \equiv 0 \qquad [2.16]$$

which means that a compression wave will be created. Furthermore it is noticed that 2.15 has a general solution in the form:

$$u = G(x - ct) + H(x + ct) \qquad [2.17]$$

where G and H are arbitrary functions. The solution thus consists of two similarity solutions, one (G) that moves in the positive x-direction and another (H) that moves in the negative x-direction. The general conclusion which may be drawn from this is perhaps better exhibited if 2.15 is expressed in terms of the stresses. For the case under consideration one has (in view of 1.33 and 1.34):

$$\left.\begin{aligned} &\sigma_x = (2G + \lambda)\frac{\partial u}{\partial x}, \quad \sigma_y = \lambda\frac{\partial u}{\partial x}, \quad \sigma_z = \lambda\frac{\partial u}{\partial x} \\ &\tau_{xy} = \tau_{xz} = \tau_{yz} = 0 \end{aligned}\right| \qquad [2.18]$$

Thus, by differentiating 2.15 once with respect to x, it may be reformulated as:

$$\frac{\partial^2\sigma_x}{\partial x^2} = \frac{1}{c^2}\frac{\partial^2\sigma_x}{\partial t^2} \qquad [2.19]$$

which is identical to 2.15 and therefore has the general solution:

$$\sigma_x = G(x - ct) + H(x + ct) \qquad [2.20]$$

It is a matter of rather routine logic to conclude from this that *if a stress pulse σ of any time dependence is applied to the free surface, this stress pulse is transmitted throughout the halfspace in its original shape (no dispersion).*

After these initial remarks we return to the problem at hand and solve it by means of the Laplace transform technique, which will subsequently be applied to most problems. Eq. 2.15 is to be solved with the following boundary and initial conditions:

$$\left.\begin{array}{lll} \sigma_x = 0 & \text{at } x \to \infty & \text{for all } t \\ \sigma_x = \sigma_0 & \text{at } x = 0 & \text{for } t > 0 \\ \sigma_x = 0 & \text{for all } x & \text{at } t = 0 \\ \dfrac{\partial u}{\partial t} = u = 0 & & \text{for } t = 0 \end{array}\right| \qquad [2.21]$$

In terms of the displacement, u, these conditions may be restated as:

$$\left.\begin{array}{ll} \dfrac{\partial u}{\partial x} = 0 & \text{for } x \to \infty \\ \dfrac{\partial u}{\partial x} = \dfrac{\sigma_0}{2G + \lambda} & \text{for } x = 0 \quad \text{and} \quad t > 0 \\ \dfrac{\partial u}{\partial x} = 0 & \text{for } t = 0 \quad \text{and for } x \to \infty \\ \dfrac{\partial u}{\partial t} = u = 0 & \text{for } t = 0 \end{array}\right| \qquad [2.22]$$

The Laplace transform is then applied in the form:

$$u^* = \mathcal{L}\{u\} = \int_0^\infty u e^{-st}\,dt \qquad [2.23]$$

The corresponding inversion formula is:

$$u = \frac{1}{2\pi i} \int_{c-i\infty}^{c+i\infty} u^* e^{+st} \, ds \qquad [2.24]$$

and will be used whenever an immediate solution cannot be found in the tables.

Application of 2.23 to 2.15 will render:

$$\frac{\partial^2 u^*}{\partial x^2} = \frac{1}{c^2} u^* s^2 \qquad [2.25]$$

where the initial conditions in 2.21 or 2.22 have been satisfied. The corresponding boundary conditions are obtained from 2.22 by application of 2.23:

$$\left.\begin{aligned} \frac{\partial u^*}{\partial x} &= 0 && \text{for } x \to \infty \\ \frac{\partial u^*}{\partial x} &= \frac{\sigma_0}{2G+\lambda} \cdot \frac{1}{s} && \text{for } x = 0 \end{aligned}\right| \qquad [2.26]$$

The solution of 2.25, which satisfies 2.26 will be:

$$\left.\begin{aligned} u^* &= -\frac{\sigma_0 c}{(2G+\lambda)} \frac{1}{s^2} e^{-sx/c} \\ \frac{\partial u^*}{\partial x} &= \frac{\sigma_0}{(2G+\lambda)} \frac{1}{s} e^{-sx/c} \end{aligned}\right| \qquad [2.27]$$

In this case the solution for u is found directly from the tables in Batemann (1954) and it can be formulated as follows:

$$u(x, t) = -\frac{\sigma_0 c}{(2G+\lambda)} \cdot \begin{cases} 0 & \text{for } 0 \leqslant t \leqslant \frac{x}{c} \\ \left(t - \frac{x}{c}\right) & \text{for } \frac{x}{c} \leqslant t \leqslant \infty \end{cases} \qquad [2.28]$$

and:

$$\sigma_x = \sigma_0 \cdot \begin{cases} 0 & \text{for } 0 \leqslant t \leqslant \frac{x}{c} \\ 1 & \text{for } \frac{x}{c} \leqslant t \leqslant \infty \end{cases} \qquad [2.29]$$

The solution is given in Fig. 5 where the stress σ_x in the solid as well as the displacement u has been plotted for two different times (t_1 and t_2). It is seen that, as time

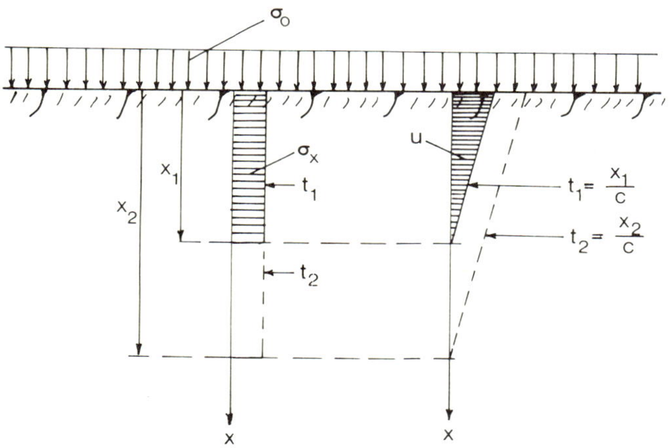

Fig. 5. Stress- and velocity distribution in an elastic halfspace at different times, t_1 and t_2 (plane wave case).

increases, the surface continues to move in the x-direction as more and more of the halfspace is being compressed. It should be noticed that the front of the wave moves with speed c of the compression wave. Finally attention is drawn to the fact that, even though the wave is here given by σ_x only, it must be remembered that

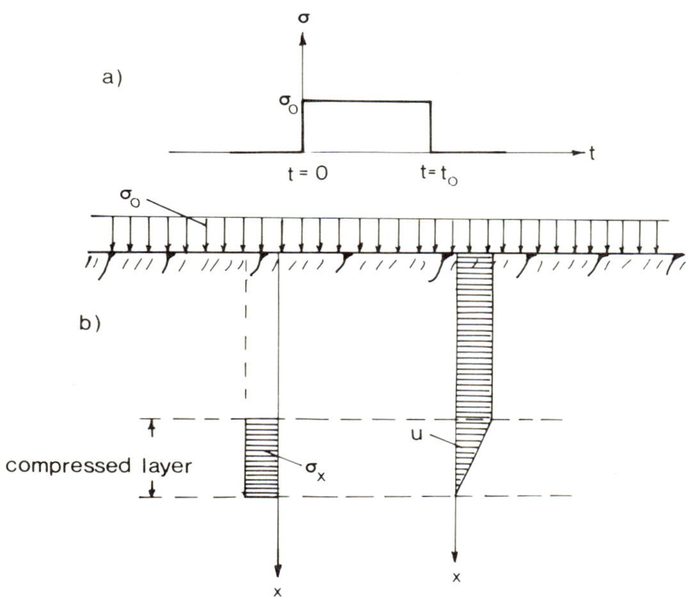

Fig. 6. Stress- and velocity distribution in an elastic halfspace at a given instant (rectangular plane wave).

the state of stress in the wave is given by 2.18 and that consequently σ_y and σ_z are equal and proportional to σ_x.

The case when stress σ_0 applied to the surface is suddenly removed again after a time t_0 may be conceived of as the difference between two Heaviside functions which thus render a rectangular input stresspulse at the surface. This case is thus immediately solved by subtracting from the present solution $G_x(t)$ the solution $G_x(t - t_0)$. The situation will then be as sketched in Fig. 6, which shows the situation at a given time. It is seen that a layer of the material is compressed and that this region will move away from the free surface with the speed of the compression wave (c). The free surface undergoes a finite displacement, which is equal to the change in thickness of the compressed layer due to the compression. It is also observed that the wave does not change its shape (no dispersion) as it moves, a fact that corresponds to the general statement made previously.

So far only the compression wave in the halfspace has been considered. It is created by the application of a normal stress to the free surface. If, however, a shear stress τ_0 is suddenly applied to the free surface, a situation is created whereby only component v of the displacement vector $\vec{s}$ will be different from zero. In this case, one concludes from 2.1 that:

$$\frac{\partial^2 v}{\partial x^2} = \frac{1}{c_t^2}\frac{\partial^2 v}{\partial t^2}$$

$$\frac{\partial v}{\partial y} = \frac{\partial v}{\partial z} = u = w \equiv 0 \qquad [2.30]$$

This gives a situation in which the state of stress in view of 1.33 and 1.34 will be given by:

$$\sigma_x = \sigma_y = \sigma_z = \tau_{yz} = \tau_{zx} = 0$$

$$\tau_{xy} \neq 0 \qquad [2.31]$$

It is furthermore clear that the state of deformation is characterized by:

$$\operatorname{div} \vec{s} = 0 \qquad [2.32]$$

and the problem at hand is that of an equivoluminal wave or a shear wave being transmitted with velocity c_t. One may now adopt the same boundary and initial conditions for v as the ones used for u in the case of the compression wave. Then 2.30 with these conditions represent the identical mathematical problem as 2.15 with its conditions, and the same solution applies in this case as in the previous one. The only difference will be that the speed of propagation is now c_t.

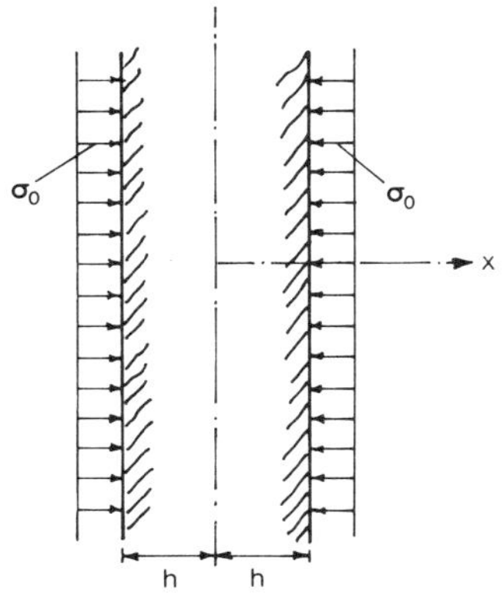

Fig. 7. The slab with an applied stress σ_0 at the surfaces.

PLANE WAVES IN A FINITE MEDIUM

Assume an infinite slab bounded by two parallel planes normal to the x-axis and a distance $2h$ apart as shown in Fig. 7. The origin of the coordinate system is located in the middle of the slab, and it is assumed that on the surface of the slab a normal stress σ_0 (evenly distributed over the surface) is suddenly applied at time $t = 0$. This stress is supposed to remain constant for $t > 0$, and the input pulse is thus again a Heaviside function in time. The problem is now to determine what happens for $t > 0$.

Again the same reasoning applies here as in the case of an infinite halfspace; i.e. $v = w = 0$, $u \neq 0$ and div $\vec{s} \neq 0$, whereas curl $\vec{s} = 0$. Thus again the case of a compression wave is at hand, and eq. 2.1 gives the governing equation in the form:

$$\frac{\partial^2 u}{\partial x^2} = \frac{1}{c^2}\frac{\partial^2 u}{\partial t^2} \qquad [2.33]$$

which is identical to 2.15. The difference between this case and the one treated previously lies in the boundary conditions. These may be stated as:

$$\left.\begin{aligned} &\sigma_x = \sigma_0 && \text{at } x = h \text{ for all } t \\ &\sigma_x = \sigma_0 && \text{at } x = -h \text{ for all } t \\ &\sigma_x = 0 && \text{for } t = 0 \text{ and all } x \\ &\frac{\partial \sigma_x}{\partial t} = 0 && \text{for } t = 0 \text{ and all } x \end{aligned}\right| \qquad [2.34]$$

where the last two statements represent the initial conditions. Because the boundary conditions are expressed in terms of stresses, it seems appropriate to differentiate 2.33 with respect to x and then use 2.18, which holds true for this case also, to convert the equation into:

$$\frac{\partial^2 \sigma_x}{\partial x^2} = \frac{1}{c^2}\frac{\partial^2 \sigma_x}{\partial t^2} \qquad [2.35]$$

The following non-dimensional quantities are now defined:

$$\left.\begin{aligned} &\text{Non-dimensional distance:} && \xi = \frac{x}{h} \\ &\text{Non-dimensional time:} && \tau = \frac{ct}{h} \\ &\text{Non-dimensional displacement:} && \Phi = \frac{2G+\lambda}{\sigma_0 h}\,u \\ &\text{Non-dimensional stress:} && S_x = \frac{\sigma_x}{\sigma_0} \end{aligned}\right| \qquad [2.36]$$

With these quantities relation 2.18 may be rewritten as:

$$S_x = S = \frac{\partial \Phi}{\partial \xi}\ \left(S_y = \frac{\lambda}{2G+\lambda}\frac{\partial \Phi}{\partial \xi}, S_z = \frac{\lambda}{2G+\lambda}\frac{\partial \Phi}{\partial \xi}\right) \qquad [2.37]$$

and the mathematical formulation is given by the differential equation:

$$\frac{\partial^2 S}{\partial \xi^2} = \frac{\partial^2 S}{\partial \tau^2} \qquad [2.38]$$

with the boundary and initial conditions:

$$\left.\begin{aligned} &S = 1 && \text{at } \xi = +1 \text{ for all } \tau \\ &S = 1 && \text{at } \xi = -1 \text{ for all } \tau \\ &S = 0 && \text{for } \tau = 0 \text{ and all } \xi \\ &\frac{\partial S}{\partial \tau} = 0 && \text{for } \tau = 0 \text{ and all } \xi \end{aligned}\right| \qquad [2.39]$$

The Laplace transform is now reformulated and will be expressed as:

$$\mathcal{L}\{S\} = S^*(\xi, s) = \int_0^\infty S(\xi, \tau)\, e^{-s\tau}\, d\tau \qquad [2.40]$$

Application of the Laplace transform to the differential equation [2.38] and its boundary conditions renders:

$$\frac{\partial^2 S^*}{\partial \xi^2} = s^2 S^* \qquad [2.41]$$

with the boundary conditions:

$$\left.\begin{aligned} S^*(1,\ s) &= \frac{1}{s} \\ S^*(-1, s) &= \frac{1}{s} \end{aligned}\right| \qquad [2.42]$$

The solution is then easily established:

$$S^* = \frac{1}{s}\frac{\cosh(\xi s)}{\cosh(s)} \qquad [2.43]$$

One might now use the Laplace inversion integral to establish the final result, but it may also be reproduced from Batemann's (1954) tables whereby:

$$S = 1 + \frac{2}{\pi}\sum_{n=1}^{n=\infty}\frac{(-1)^n}{(n-\frac{1}{2})}\cos\left[(n-\tfrac{1}{2})\,\pi\xi\right]\cos\left[(n-\tfrac{1}{2})\,\pi\tau\right] \qquad [2.44]$$

$$\Phi = \xi + \frac{2}{\pi^2}\sum_{n=1}^{n=\infty}\frac{(-1)^n}{(n-\frac{1}{2})^2}\sin\left[(n-\tfrac{1}{2})\,\pi\xi\right]\cos\left[(n-\tfrac{1}{2})\,\pi\tau\right] \qquad [2.45]$$

Attention is here drawn to the fact that τ is defined such that $\tau = 1$ is the time needed for the front of the stress wave to travel half way through the slab.

Consequently, the solution for S is plotted in Fig. 8 for $0 \leqslant \tau \leqslant 4$. It is seen how the plane wave starts out from each surface of the slab as in the case of an infinite medium. ("The wave 'does not know' yet that the medium is bounded"). At time $\tau = 1$, the two shock fronts meet in the middle of the slab. They travel, however, undisturbed on, and in the region between the two fronts there will be an increasing portion of the slab in which the stresses from the two waves are superimposed so that in this region $S = 2$, i.e. twice as great as the stress applied to the surfaces. At $\tau = 2$, the entire slab has a stress $S = 2$. What happens in the subsequent period $2 \leqslant \tau \leqslant 4$ can be interpreted as a tension wave starting out from both surfaces and superimposing itself on the already existing state of stress in the slab. At $\tau = 3$, the fronts of the tension waves meet in the middle of the slab and continue to move undisturbed. At $\tau = 4$, the fronts of the tension waves have reached the opposite surfaces, and the entire slab is released (no stress). But, in the same way as the compression waves were reflected at the free surfaces and came back as tension waves, the tension waves are also reflected at the free surfaces and come back as compression waves. Thus at $\tau = 4$, the whole process starts over again as when $\tau = 0$, and the slab will oscillate forever as shown, with a period $\tau = 4$. Thus, the static case where the outside stress, σ_0, creates a static stress equal to σ_0 throughout the thickness of the slab is never reached.

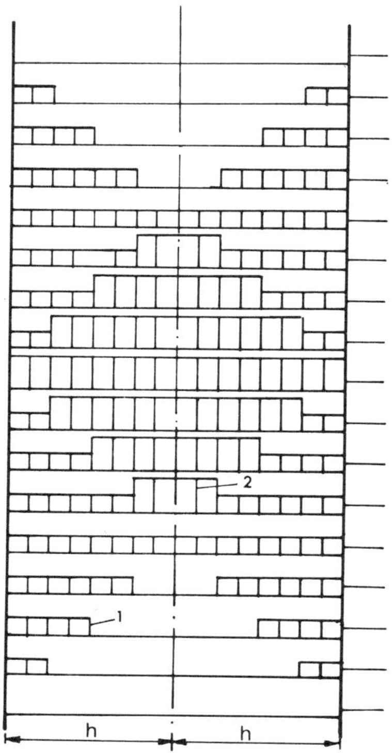

Fig. 8. The stress distribution in the slab at consecutive instances (Heaviside input pulse).

So far the example treated here has drawn attention to two very important features of the propagation of stress waves: (1) the reflection of a stress wave from a free surface, and (2) the absence of a damping effect in the ideally elastic medium.

It is clear that the second feature will be of importance when the theory is applied to the propagation of stress waves in real materials. It is, however, also clear that certain features of the propagation of stress waves in an elastic medium may be more or less characteristic for the wave propagation in real media under certain circumstances. It is the object of the present text to point out such cases, one of them being the reflection of a wave at a free surface.

THE NORMAL REFLECTION OF A STRESS WAVE FROM A FREE PLANE SURFACE

The reflection of a wave from a free plane surface has to some extent been demonstrated in Fig. 8. However, the situation described in the preceding paragraph is obscured by the interference of two waves and thus does not bring out clearly the

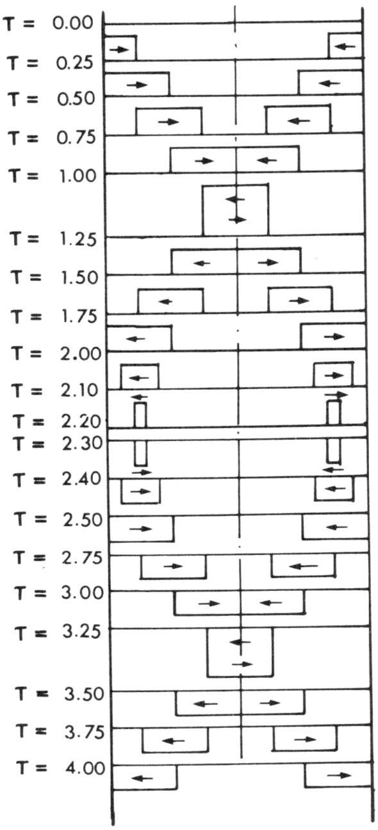

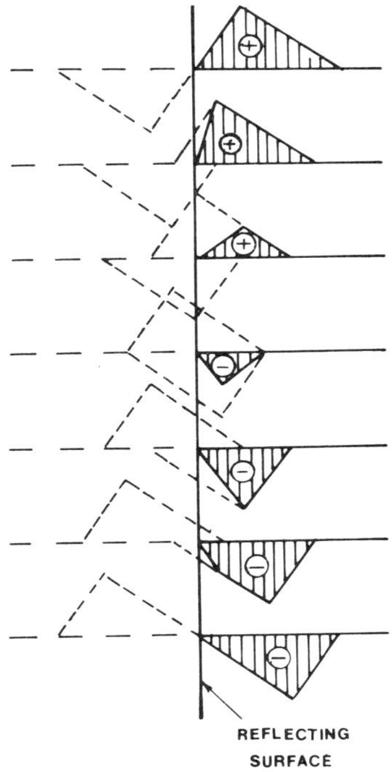

Fig. 9. The stress distribution in the slab at consecutive instances (rectangular input pulse). Normal reflection from a plane surface.

Fig. 10. Normal reflection of a plane stress wave from a plane surface (triangular input pulse).

essential features of the reflection. The solution obtained permits, however, the solution for a rectangular wave to be set up without further calculations. If stress σ_0, applied to the surfaces of the slab at $\tau = 0$ in Fig. 7, is suddenly removed at $\tau = \tau_0$, a rectangular input pulse is created with the duration τ_0. The solution will be the same as before for $\tau < \tau_0$. At $\tau = \tau_0$, a tension wave is added to the solution of the compression wave. The solution will then be the same as for the compression wave with a shift of τ_0 in time. Thus the situation will be as sketched in Fig. 9 where $\tau_0 = 0.5$ has been assumed to make the pulse so short that no interference between the two pulses occurs during the reflection period. At $\tau = 1.0$ the two pulses meet at the center of the slab. They travel undisturbed on, and because they are superimposed on each other, the stress at the centerline will be $S = 2$ for $1.0 < \tau$

< 1.5. At $\tau = 2.00$ the wave front reaches the free surface and the reflection starts. This reflection takes place when $2.0 \leqslant \tau \leqslant 2.5$ and the stress at any point may be conceived of as a superposition of the compression wave and its negative "mirror image" (tension wave) that moves in the opposite direction, always with the free plane surface as the mirror. Thus at $\tau = 2.50$, a tension wave is leaving the free surface and travelling in the opposite direction of the originally oncoming compression wave. The two reflected tension waves now move as shown and another reflection period begins at $\tau = 4$. Then the whole thing repeats itself, and the effect of the original input pulse will be that pulses travel back and forth in the slab forever as shown in Fig. 9 for the first period. It is clear that this is unacceptable from a physical point of view when attempts are made to apply the theory to real materials. However, the reflection process stated above does indeed describe what happens in a real material. To make the description even clearer, the reflection of a triangular wave is shown in its different stages in Fig. 10. The two waves are shown with dashed lines and the real stress situation is shown shaded. Otherwise the figure is thought to be self explanatory.

PARTICLE VELOCITY AND ITS REFLECTION AT A FREE PLANE SURFACE

So far the stress wave has been discussed in terms of the spatial stress distribution in the wave. However, as long as the wave is plane, the general solution [2.17] is valid, and it may be written as:

$$u = G(\eta_+) + H(\eta_-) \qquad \text{with } \eta_+ = x - ct,\ \eta_- = x + ct \tag{2.46}$$

Therefore, considering only a wave which moves in the positive x-direction, one has:

$$\frac{\partial u}{\partial x} = \frac{\mathrm{d}G}{\mathrm{d}\eta_+}, \quad \frac{\partial u}{\partial t} = -\frac{\mathrm{d}G}{\mathrm{d}\eta_+}c \tag{2.47}$$

Here $\partial u/\partial t$ is the particle velocity, and in view of 2.18, eqs. 2.47 will give:

$$\frac{\sigma_x}{(2G + \lambda)} = -\frac{1}{c}\frac{\partial u}{\partial t} \tag{2.48}$$

This means that there is a direct linear relationship between the stress, σ_x, and the particle velocity, $\partial u/\partial t$, in a plane wave. Consequently the distribution of stress in the wave is also a distribution of the particle velocity in the wave.

If now the wave moves in the negative x-direction, the situation is changed somewhat because then it is described by the function H in 2.46, i.e.:

$$\left.\begin{aligned}\frac{\partial u}{\partial x} &= \frac{\mathrm{d}H}{\mathrm{d}\eta_-}\\ \frac{\partial u}{\partial t} &= +c\,\frac{\mathrm{d}H}{\mathrm{d}\eta_-}\end{aligned}\right\}\quad \frac{\sigma_x}{2G+\lambda} = \frac{1}{c}\,\frac{\partial u}{\partial t} \qquad [2.49]$$

Thus:

(1) A compression wave moving in the negative x-direction and a tension wave moving in the positive x-direction create negative particle velocities.

(2) A compression wave moving in the positive x-direction and a tension wave moving in the negative x-direction create positive particle velocities.

This result is often used in cases when the wave is not plane, but conditions are

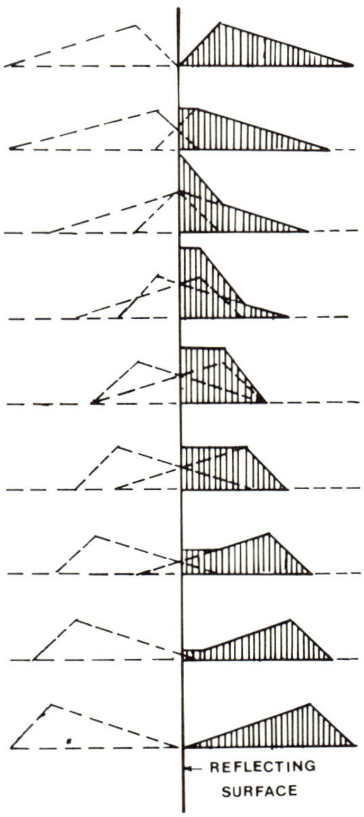

Fig. 11. Normal reflection from a plane surface of a velocity wave (triangular input pulse).

such that the plane wave seems a reasonable approximation. Such cases from measurements in rocks will be presented.

The considerations leading to the presented result have a bearing on the reflection process of a wave at a plane free surface. Reconsidering Fig. 10 it is found that the oncoming wave is a compression wave, denoted by +, moving in the negative x-direction, whereas its negative "mirror image" is a tension wave, denoted by –, moving in the positive x-direction. Thus, according to case 1 above, the velocity fields associated with both waves are of the same sign and the superposition of the velocity fields will be as described in Fig. 11. It is first of all noticed that the reflected wave (now a velocity wave) is of the same sign as the oncoming wave. The reflection is described as a superposition of the wave on its positive "mirror image".

It is observed that the maximum velocity measured on the free surface is two times the maximum particle velocity in the wave. This is important in cases where measuring devices are mounted on free surfaces as the case very often is when measurements are taken in rocks.

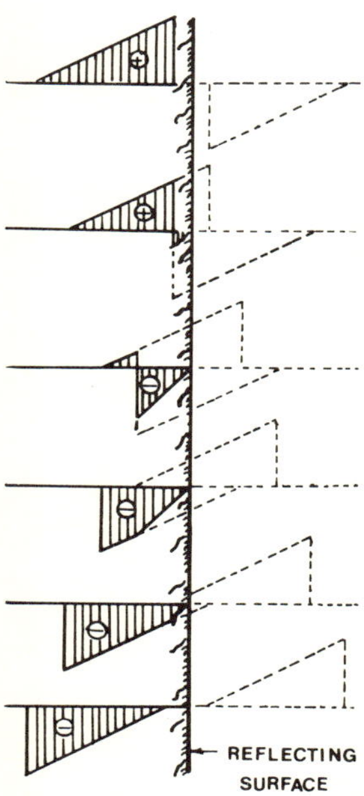

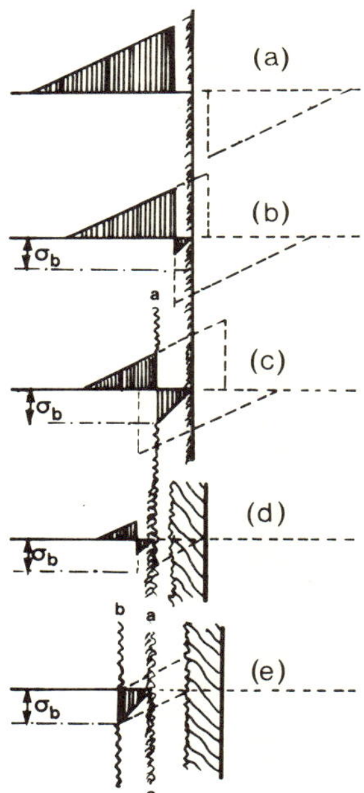

Fig. 12. Normal reflection of a triangular wave from a plane surface.

Fig. 13. Spalling at a plane surface.

SPALLING

Very often it is observed that a stress wave, when reflected from a surface, will cause some part of the material close to the surface to be "ripped off". This is called "spalling", and the mechanism of spalling may be understood with reference to the reflection process. Spalling is almost exclusively observed in materials which exhibit greater strength in compression than in tension (examples are: cast iron, concrete, rock). Fig. 12 shows the reflection of a triangular compression wave in such a material under the assumption that the maximum stress amplitude in the wave is less than the tensile stress, σ_b, for which fracture occurs. Fig. 13 shows the same reflection under the assumption that the maximum amplitude of the oncoming compression wave is equal to $-2\sigma_b$. Stage (a) and (b) are identical to the corresponding stages in Fig. 12. Stage (c), however, shows the situation when the maximum amplitude of the reflected tension wave is equal to σ_b. Fracture occurs then along a–a, and the slab, now loose from the rest of the material, moves away as indicated in stage (d). That it really does move as shown can be deduced from the field of particle velocities. Stage (c) represents now the start of a reflection procedure from plane a–a for the remaining part of the compression wave. Stage (e) shows the situation in stage (c) repeating itself. A new fracture will occur along b–b. But because the maximum amplitude of the wave was chosen equal to $-2\sigma_b$, no part of the compression wave remains and no tension wave is reflected. A double spalling has occurred.

Chapter 3

SOME CHARACTERISTIC FEATURES OF SPHERICAL WAVES (ELASTIC MEDIUM)

INTRODUCTION

The scope of this text being the study of the wave propagation in rocks makes it important to concentrate on what happens close to a detonating charge. Detonating charges are for the most part the origin of stress waves, and whether these charges are aimed purely at creating stress waves for observation in connection with experimental investigations or used for blasting purposes, they are mostly well-confined charges, which may be studied theoretically by means of stress waves with a spherical symmetry. This gives the general background for the present chapter.

BASIC EQUATIONS

While the basic equations were deduced in Cartesian coordinates, the case of stress waves with a spherical symmetry is most efficiently studied in a spherical coordinate system. It will be assumed that the geometry of the problem (Fig. 14) is such that the explosion can be replaced by a pressure p acting on the inside of a spherical cavity in an infinite medium. The pressure is supposed to be evenly distributed over the inside surface of the cavity and it is assumed that pressure p is a

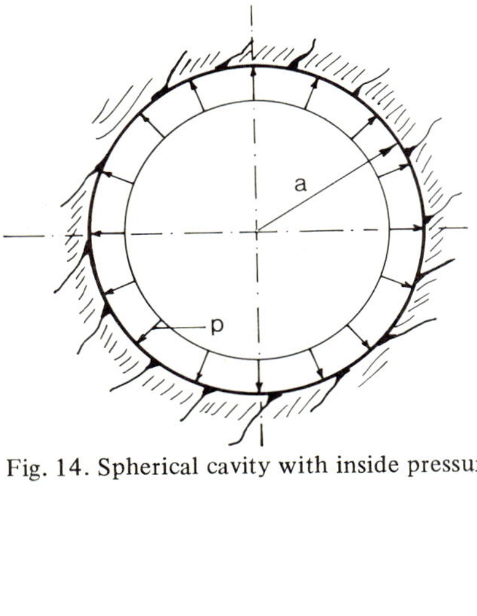

Fig. 14. Spherical cavity with inside pressure as origin of a spherical wave.

known function of time, i.e.:

$$p = p_0 f(t) \qquad [3.1]$$

It is furthermore assumed that the maximum value of $f(t)$ is equal to 1, i.e. p_0 is the maximum value of p. Under these circumstances one may conclude that at all times will spherical symmetry prevail, i.e. all quantities will be functions of the distance, r, from the center of the cavity and of time t. Furthermore, the displacement vector, $\vec{s}$, will have only one component:

$$\vec{s} = \vec{i}_r v_r \qquad [3.2]$$

Under these circumstances one obtains from eqs. 1.51, 1.52 and 1.53:

Strain:

$$\epsilon_r = \frac{\partial v_r}{\partial r}, \quad \epsilon_\varphi = \frac{v_r}{r}, \quad \epsilon_\theta = \frac{v_r}{r} \qquad [3.3]$$

Shear strain:

$$\gamma_{r\varphi} = \gamma_{\varphi\theta} = \gamma_{\theta r} = 0 \qquad [3.4]$$

Equation of motion:

$$\frac{\partial \sigma_r}{\partial r} + \frac{2\sigma_r - \sigma_\varphi - \sigma_\theta}{r} = \rho \frac{\partial^2 v_r}{\partial t^2} \qquad [3.5]$$

The relation between the state of stress and the state of deformation is given in 1.29 which in this simplified case will give:

$$I_1 = \sigma_r + \sigma_\varphi + \sigma_\theta = (2G + 3\lambda)\left(\frac{\partial v_r}{\partial r} + \frac{2v_r}{r}\right) \qquad [3.6]$$

$$\left.\begin{aligned} \sigma_r - \tfrac{1}{3}I_1 &= 2G\left[\frac{\partial v_r}{\partial r} - \frac{1}{3}\left(\frac{\partial v_r}{\partial r} + \frac{2v_r}{r}\right)\right] \\ \sigma_\varphi - \tfrac{1}{3}I_1 &= 2G\left[\frac{v_r}{r} - \frac{1}{3}\left(\frac{\partial v_r}{\partial r} + \frac{2v_r}{r}\right)\right] \\ \sigma_\theta - \tfrac{1}{3}I_1 &= 2G\left[\frac{v_r}{r} - \frac{1}{3}\left(\frac{\partial v_r}{\partial r} + \frac{2v_r}{r}\right)\right] \end{aligned}\right| \qquad [3.7]$$

It is a matter of routine algebra to deduce the expressions for σ_r, σ_φ and σ_θ which, introduced into 3.5 will give the following differential equation for v_r:

$$\frac{\partial^2 v_r}{\partial r^2} + \frac{2}{r}\frac{\partial v_r}{\partial r} - \frac{2}{r^2} v_r = \frac{1}{c^2}\frac{\partial^2 v_r}{\partial t^2} \qquad [3.8]$$

The non-dimensional quantities may be introduced as follows:

$$\begin{aligned} &\text{Non-dimensional distance:} && \xi = \frac{r}{a} \\ &\text{Non-dimensional time:} && \tau = \frac{tc}{a} \\ &\text{Non-dimensional displacement:} && \Phi = \frac{E v_r}{a(1+\nu)\,p_0} \\ &\text{Non-dimensional stress:} && S_r = \frac{\sigma_r}{p_0} \end{aligned} \qquad [3.9]$$

With these quantities the differential equation [3.8] will be:

$$\frac{\partial^2 \Phi}{\partial \xi^2} + \frac{2}{\xi}\frac{\partial \Phi}{\partial \xi} - \frac{2}{\xi^2}\Phi = \frac{\partial^2 \Phi}{\partial \tau^2} \qquad [3.10]$$

The boundary and initial conditions are given in Table I.

Because the conditions are known in terms of stresses, whereas the differential equation is stated in terms of the displacement (Φ), the following relation should be

TABLE I

Boundary and initial conditions

For:	Physical conditions	For:	Non-dimensional	No.
$\sigma_r(r,t)$	$\sigma_r(a,t) = p_0 f(t)$	$S_r(\xi,\tau)$	$S_r(1,\tau) = f(a\tau/c)$	(1)
	$\sigma_r(\infty,t) = 0$		$S_r(\infty,\tau) = 0$	(2)
	$\sigma_r(r,0) = 0$		$S_r(\xi,0) = 0$	(3)
	$\left.\frac{\partial \sigma_r}{\partial t}\right\|_{t=0} = 0$		$\left.\frac{\partial S_r}{\partial \tau}\right\|_{\tau=0} = 0$	(4)
$v_r(r,t)$	$v_r(r,0) = 0$	$\Phi(\xi,\tau)$	$\Phi(\xi,0) = 0$	(5)
	$\left.\frac{\partial v_r}{\partial t}\right\|_{t=0} = 0$		$\left.\frac{\partial \Phi}{\partial \tau}\right\|_{\tau=0} = 0$	(6)

kept in mind:

$$S_r = \frac{\partial \Phi}{\partial \xi} + \frac{\nu}{1-2\nu}\left(\frac{\partial \Phi}{\partial \xi} + \frac{2\Phi}{\xi}\right) \tag{3.11}$$

where relation 1.31 has been used.

The solution to the problem is now deduced by means of the Laplace transform as defined in 2.40. Applied to the differential [3.10] the following equation is obtained:

$$\frac{\partial^2 \Phi^*}{\partial \xi^2} + \frac{2}{\xi}\frac{\partial \Phi^*}{\partial \xi} - \frac{2}{\xi^2}\Phi^* = s^2 \Phi^* \tag{3.12}$$

This is the differential equation for $\Phi^*(\xi,s)$, where s appears as a parameter only. The solution to this equation may be written in terms of the modified Bessel functions $I_\nu(z)$ and $K_\nu(z)$ as defined in Batemann (1953):

$$\Phi^*(\xi, s) = [A\, I_{3/2}(s\xi) + BK_{3/2}(s\xi)]\ \xi^{-1/2} \tag{3.13}$$

Here A and B are constants of integration, which will depend on the Laplace parameter s. They are determined by boundary conditions 1 and 2 in Table I. These will upon application of the Laplace transform be:

$$S_r^*(1, s) = f^*(s), \quad S_r^*(\infty, s) = 0 \tag{3.14}$$

Eq. 3.11 gives the last relation necessary for a complete statement of the problem:

$$S_r^* = \frac{\partial \Phi^*}{\partial \xi} + \frac{\nu}{1-2\nu}\left(\frac{\partial \Phi^*}{\partial \xi} + \frac{2\Phi^*}{\xi}\right) \tag{3.15}$$

It is concluded from 3.14 and 3.15 that the conditions at infinity will be satisfied if $\partial\Phi^*/\partial\xi$ and Φ^* vanish as $\xi \to \infty$. From the definition of the modified Bessel functions it is found that this will be the case if the constant of integration $A = 0$. Furthermore, the modified Bessel function of order 3/2 may be expressed in terms of elementary functions as follows:

$$K_{3/2}(z) = \sqrt{\frac{\pi}{2}}\, z^{-1/2}\, e^{-z}\left(1 + \frac{1}{z}\right) \tag{3.16}$$

Thus, from eq. 3.13:

$$\Phi^*(\xi, s) = B\sqrt{\frac{\pi}{2}}\,(s\xi)^{-1/2}\, e^{-s\xi}\left(1 + \frac{1}{s\xi}\right)\xi^{-1/2} \tag{3.17}$$

and from eq. 3.15:

$$S_r^* = -B\sqrt{\frac{\pi}{2}}\,s^{3/2}\,[2(s\xi)^{-3} + 2(s\xi)^{-2} + A(\nu)\,(s\xi)^{-1}]\,e^{-s\xi} \qquad [3.18]$$

where:

$$A(\nu) = \frac{1-\nu}{1-2\nu} \qquad [3.19]$$

The solution presented in 3.18 satisfies boundary condition 2 in Table I, and initial conditions 3 and 4. The only remaining condition is 3.14 which gives:

$$f^*(s) = -B\sqrt{\frac{\pi}{2}}\,s^{3/2}\,[2s^{-3} + 2s^{-2} + A(\nu)\,s^{-1}]\,e^{-s} \qquad [3.20]$$

This determines B and finally the solution (eq. 3.18) will be:

$$S_r^* = \frac{A(\nu)\,(s\xi)^2 + 2(s\xi) + 2}{A(\nu)\,s^2 + 2s + 2}\cdot\frac{f^*(s)}{\xi^3}\,e^{-s(\xi-1)} \qquad [3.21]$$

This solution is general in the sense that $f^*(s)$ has not been specified yet. This means that the solution is valid for any input pulse $f(t)$. For each case the input pulse must be specified and the inverse Laplace transform must be computed.

THE HEAVISIDE FUNCTION AS INPUT PULSE

The first example to be considered here is the case when the pressure, p, is impulsively applied at $t=0$ and then kept at the constant value p_0 at all times. Thus:

$$p = p_0 f(t) = \begin{cases} p_0 & t > 0 \\ 0 & t < 0 \end{cases} \qquad [3.22]$$

The Laplace transform of this function will be:

$$f^*(s) = \frac{1}{s} \qquad [3.23]$$

whereupon the solution (eq. 3.21) becomes:

$$S_r^* = \frac{1}{s}\cdot\frac{A(\nu)\,(s\xi)^2 + 2(s\xi) + 2}{A(\nu)\,s^2 + 2s + 2}\cdot\frac{1}{\xi^3}\,e^{-s(\xi-1)} \qquad [3.24]$$

The inverse Laplace transform may now formally be given as:

$$S_r(\xi, \tau) = \frac{\xi^{-3}}{2\pi i} \int_{b-i\infty}^{b+i\infty} \frac{1}{s} \cdot \frac{A(\nu)(s\xi)^2 + 2(s\xi) + 2}{A(\nu)s^2 + 2s + 2} e^{-s(\xi-1-\tau)} ds \qquad [3.25]$$

Here s is conceived of as the complex variable in the s-plane, and b is to be chosen such that all singularities of the integrand are to be found to the left of the line of integration from $b - i\infty$ to $b + i\infty$.

The integrand will have its three singularities at $s = s_0$, $s = s_1$ and $s = s_2$ where:

$$\left.\begin{aligned} s_0 &= 0 \\ s_1 &= \frac{1}{A(\nu)}\left[-1 + i\sqrt{2A(\nu) - 1}\right] \\ s_2 &= \frac{1}{A(\nu)}\left[-1 - i\sqrt{2A(\nu) - 1}\right] \end{aligned}\right| \qquad [3.26]$$

From the expression for $A(\nu)$ in 3.19 it can be concluded that the two poles at s_1 and s_2 will for $0 < \nu < 0.5$ always be located in the negative (left-hand) half plane of the s-plane.

When evaluating the integral in eq. 3.25 it is observed that for $\xi - 1 - \tau > 0$ the integrand will vanish for $|s| \to \infty$ as long as $Re(s) > 0$. This means that the path of integration may be changed over to the "big half circle" in the right-hand half plane and the result will be:

$$S_r(\xi, \tau) = 0 \qquad \text{for } \xi - 1 > \tau \qquad [3.27]$$

The physical reality of this is that at any given instant τ, the region where $\xi - 1 > \tau$ will not yet have been influenced by the progressing compression wave, i.e. the situation is analogous to the plane wave. The wave will proceed with a shock front that moves with velocity c.

For $\xi - 1 - \tau < 0$, the path of integration is changed over to the "big half circle" in the left-hand half plane of the s-plane, and in this case the integral will be equal to the sum of the residues at the three poles of the integrand, i.e.:

$$S_r(\xi, \tau) = \xi^{-3} + \frac{e^{-T/A}}{\sqrt{2A-1}} \left[(-\xi^{-1} + 2\xi^{-2} - \xi^{-3}) \sin\left(\frac{T}{A}\sqrt{2A-1}\right) + \right.$$

$$\left. + (\xi^{-1} - \xi^{-3})\sqrt{2A-1} \cos\left(\frac{T}{A}\sqrt{2A-1}\right)\right] \qquad [3.28]$$

where:

$$A = A(\nu) = \frac{1-\nu}{1-2\nu} \qquad T = \tau - (\xi - 1) \qquad [3.29]$$

This solution is valid for $T > 0$, and thus the complete solution is known. It carries in it a number of interesting features which will be exhibited in the following.

(1) For $\tau \to \infty$, the solution will be:

$$S_r(\xi, \infty) = \xi^{-3} \qquad [3.30]$$

This is the static solution and gives the stress distribution around a spherical cavity inside which a pressure p_0 acts statically.

(2) For $\tau = \xi - 1$, $(T = 0)$ i.e. at the shock front:

$$S_r = \xi^{-1} \qquad [3.31]$$

This is the maximum amplitude of the shock wave at any point ξ. It is not dependent on ν. It is larger than the static value, and it means that at any point in the material the shock wave will create a stress S_r which, at the arrival of the front of the shock wave, impulsively reaches the value ξ^{-1} and then gradually falls off to the value ξ^{-3}.

(3) The solution (eq. 3.28) contains terms which at any point ξ can be conceived of as a damped oscillation. The frequency ω of this oscillation is:

$$\omega = \frac{c}{aA}\sqrt{2A-1} \qquad [3.32]$$

and it is independent of location, but depends on Poisson's ratio for the material. From a dimensional analysis of the problem, the introduction of the radius, a, of the cavity is equivalent with an introduction of an eigenfrequency $\omega_0 = c/a$ of the cavity. Thus the appearance of ω in the solution is not surprising. A damped oscillation in a material which does not exhibit a dissipated behavior is, however, rather unexpected.

The solution given in eqs. 3.27 and 3.28 may now be displayed in two different ways. In Fig. 15 the attention is fixed at one particular position ($\xi = 1.5$) and the variation of S_r with time (τ) is displayed. This corresponds to the signal one would obtain from a pickup located at $\xi = 1.5$ and designed to record S_r as time goes by. It is seen that in the time interval $0 < \tau < 0.5$ nothing is observed. At time $\tau = 0.5$, a sudden jump occurs and S_r reaches its maximum value, $S_{r,\max} = 0.6667$. It then

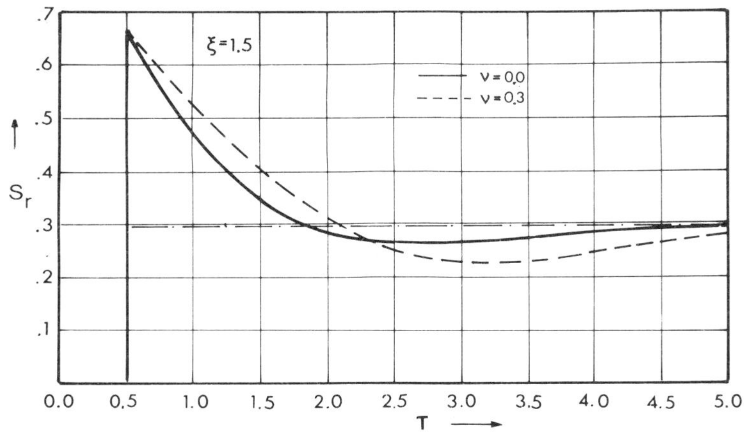

Fig. 15. Stress S as a function of time τ at the distance ξ = 1.5 for two different values of ν (Heaviside input pulse).

decreases and reaches its final value, S_r = 0.2962, after a damped oscillation. The development is displayed for $\nu = 0$ and $\nu = 0.3$ to exhibit the influence of Poisson's ratio. It is seen that basically the same development occurs in both cases.

In Fig. 16 the attention is fixed at three particular instances (τ = 1.0, 1.6 and 2.2), and the variation of S_r with location (ξ) is displayed. This corresponds to some extent to the way in which experimental results are exhibited. In each diagram two curves are drawn (— · — ·). The upper one represents ξ^{-1} and thus gives the maximum value of the shock wave at each location. The lower one represents ξ^{-3} and gives the final value of S_r for larger values of time. It is thus seen that, when the front of the shock wave is located at ξ = 3.2, the state of stress in the interval $1 \leqslant \xi \leqslant 1.5$ has almost reached its final static stage.

One more remark seems appropriate at this point. The upper curve (ξ^{-1}) represents the curve giving $S_{r,\max}$ as a function of ξ. It is seen that, in contrast to the plane wave, this case shows an attenuation of $S_{r,\max}$. This is a geometric attenuation. In many applications this curve is the essential one, because it tells the maximum effect in one place when an event takes place in another. For applications in rocks such curves become very important and experimental investigations of rocks are very often centered on experimentally finding this curve from observed maximum values measured by pickups at different locations.

THE RECTANGULAR INPUT PULSE

As in the case of the plane wave, a rectangular input pulse may be obtained by

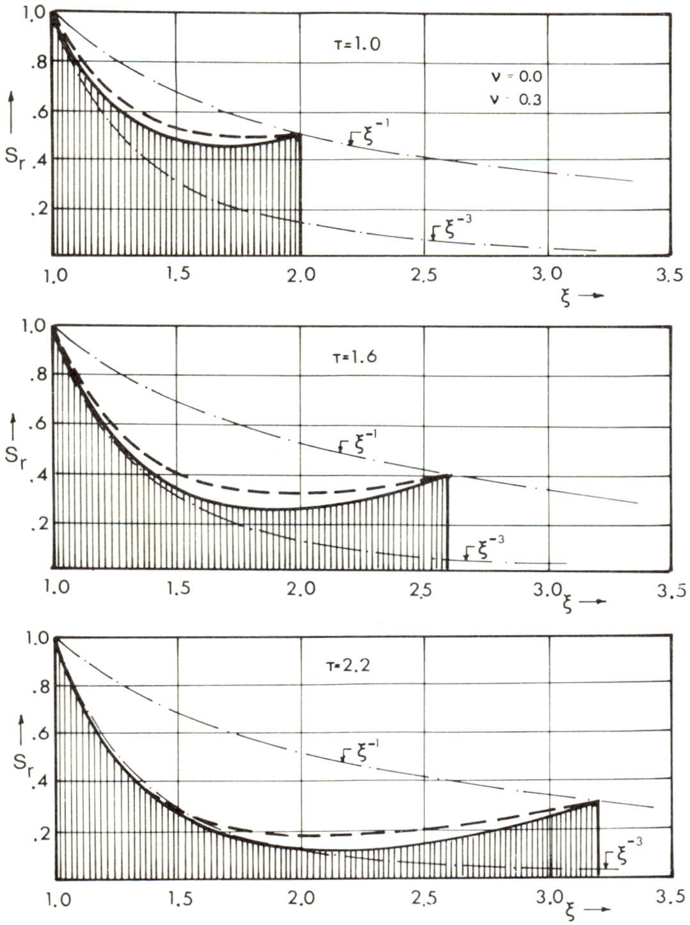

Fig. 16. Stress distribution in the material at different times, τ (Heaviside input pulse).

assuming two Heaviside input functions superimposed on each other but displaced a period τ_0 in time. The input function will then be:

$$p = \begin{cases} p_0 & 0 < t < t_0 \\ 0 & t > t_0 \end{cases}$$

or: [3.33]

$$f\left(\frac{at}{c}\right) = \begin{cases} 1 & 0 < \tau < \tau_0 \\ 0 & \tau > \tau_0 \end{cases}$$

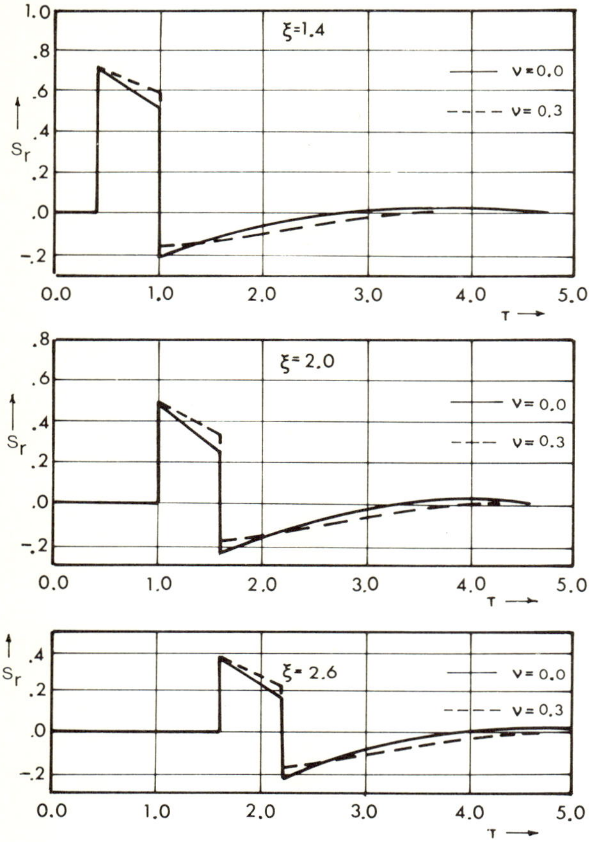

Fig. 17. Stress S as a function of time τ at different distances, ξ (rectangular input pulse). τ_0 = 0.6.

The solution to this problem is obtained through the superposition of two solutions given in the preceding section, one displaced τ_0 in time against the other. Further mathematical deductions are thus unnecessary and the results are shown in Figs. 17 and 18. Fig. 17 displays the variation with time of the stress at three locations (ξ = 1.4, 2.0 and 2.6) for a pulse of the duration $\tau_0 = 0.6$. These curves may be looked upon as pickup signals at the given locations. As before, the solid lines are valid for $\nu = 0$, the dotted lines for $\nu = 0.3$. It is again seen that a change in ν does not introduce any basic change in the signals. Fig. 18 displays the state of stress as a function of location (ξ) for three different times (τ = 1.0, 1.6, 2.2). It is seen that the stress wave consists of a region in compression which corresponds to the duration, τ_0, of the pulse. The whole region between the cavity and the compressed portion is in tension.

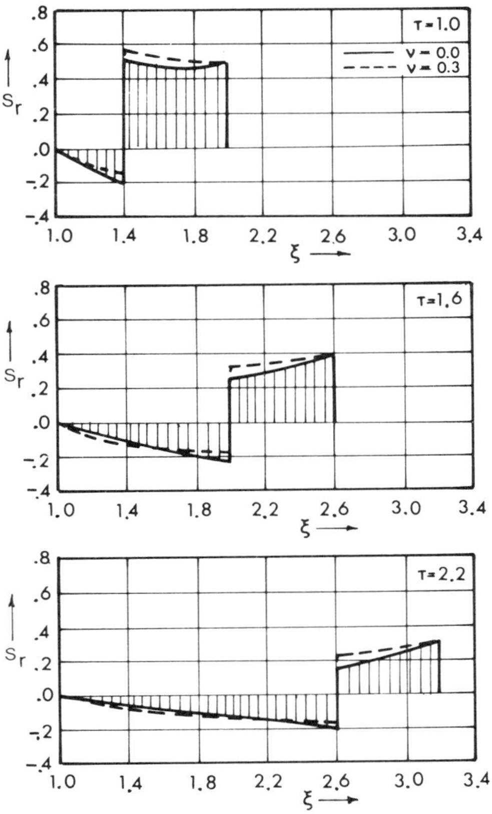

Fig. 18. Stress distribution in the material at different times, τ (rectangular input pulse). τ_0 = 0.6.

Some features of this solution should be kept in mind. First, the maximum value of the stress wave is the same as for the Heaviside pulse (ξ^{-1}). This is due to the discontinuous character of the solution. ("The material does not know that the arriving pulse is finite".) The rectangular pulse has a discontinuity both at the beginning and at the end. Such discontinuities are very strongly transmitted. Secondly, the duration of the pulse measured at any location (ξ) is the same as the duration, τ_0, at input. These features must now be investigated further, and an example will be used to exhibit the results.

THE SINUSOIDAL INPUT PULSE

The results of the preceding section drew the attention to the pulse length measured at a given location as compared with the length, τ_0, of the input pulse. The effect of the discontinuity of the input pulse was also questioned. If an input pulse is to be created experimentally, it is never possible to create such discontinuities as the ones described by a rectangular pulse. A more realistic pulse would be one which had a sinusoidal shape, and therefore the following assumption will here be made for the input pulse:

$$f(t) = \begin{cases} \sin\left(\omega_1 \dfrac{ct}{a}\right) & \text{for } 0 \leqslant t \leqslant \dfrac{\pi a}{c\omega_1} \\ 0 & \text{for } t > \dfrac{\pi a}{c\omega_1} \end{cases} \qquad [3.34]$$

Thus:

$$f\left(\frac{ct}{a}\right) = \begin{cases} \sin(\omega_1 \tau) & \text{for } 0 \leqslant \tau \leqslant \dfrac{\pi}{\omega_1} \\ 0 & \text{for } \tau > \dfrac{\pi}{\omega_1} \end{cases} \qquad [3.35]$$

giving a pulse duration $\tau_0 = \pi/\omega_1$. The Laplace transform of 3.35 renders:

$$f^* = \frac{\omega_1}{s^2 + \omega_1^2} (1 + \mathrm{e}^{-s\pi/\omega_1}) \qquad [3.36]$$

Introduced into eq. 3.21 the solution for S_r^* is obtained, whereupon S_r may formally be given in the following form:

$$S_r = \frac{\xi^{-3}}{2\pi i} \int_{b-i\infty}^{b+i\infty} \frac{[A(s\xi)^2 + 2(s\xi) + 2]\ \omega_1}{(As^2 + 2s + 2)(s^2 + \omega_1^2)} (1 + \mathrm{e}^{-s\pi/\omega_1})\, \mathrm{e}^{-s(\xi - 1 - \tau)}\, \mathrm{d}s \qquad [3.37]$$

It is seen that this expression can be divided into two parts:

$$S_r(\xi, \tau) = S_1(\xi, \tau) + S_2(\xi, \tau) \qquad [3.38]$$

where:

$$S_1(\xi, \tau) = \frac{\xi^{-3}}{2\pi i} \int_{b-i\infty}^{b+i\infty} \frac{[A(s\xi)^2 + 2(s\xi) + 2]\ \omega_1}{(As^2 + 2s + 2)(s^2 + \omega_1^2)}\, \mathrm{e}^{-s(\xi - 1 - \tau)}\, \mathrm{d}s \qquad [3.39]$$

and:

$$S_2(\xi, \tau) = S_1\left(\xi, \tau - \frac{\pi}{\omega_1}\right) \qquad [3.40]$$

By deforming the path of integration one can as before show that:

$$\begin{aligned} S_1(\xi, \tau) &= 0 \quad \text{for } \tau < \xi - 1 \\ S_2(\xi, \tau) &= 0 \quad \text{for } \tau < \xi - 1 + \frac{\pi}{\omega_1} \end{aligned} \qquad [3.41]$$

When $\tau > \xi - 1$ the solution is, as before, obtained as a sum of the residues at the poles of the integrand in 3.39, which occurs at:

$$s = s_1, \quad s = s_2, \quad s_3 = +i\omega_1, \quad s_4 = -i\omega_1 \qquad [3.42]$$

where s_1 and s_2, as before, are given in eq. 3.26. The rather tedious algebra is omitted here and the result may be stated in the following form:

$$\begin{aligned} S_1(\xi, \tau) = f_1(\xi) \sin(\omega_1 T) + f_2(\xi) \cos(\omega_1 T) + \left[f_3(\xi) \sin\left(\beta \frac{T}{A}\right) \right. \\ \left. - f_4(\xi) \cos\left(\beta \frac{T}{A}\right) \right] e^{-T/A} \end{aligned} \qquad [3.43]$$

where:

$$T = \tau - (\xi - 1), \quad \beta = \sqrt{2A - 1} \qquad [3.44]$$

$$\begin{aligned} f_1(\xi) &= \frac{(2 - A\omega_1^2 \xi^2)(2 - A\omega_1^2) + 4\omega_1^2 \xi}{\xi^3 [(2 - A\omega_1^2)^2 + 4\omega_1^2]} \\ f_2(\xi) &= \frac{2\omega_1(\xi - 1)(2 + A\omega_1^2 \xi)}{\xi^3 [(2 - A\omega_1^2)^2 + 4\omega_1^2]} \end{aligned} \qquad [3.45]$$

$$\begin{aligned} f_3(\xi) &= \frac{2\omega_1}{\xi^3 \beta N} \left[(\xi - 1)^2 \left(\frac{2}{A} + \omega_1^2\right) + (\xi^2 - 1)\left(\frac{2}{A} - \omega_1^2\right) \beta^2 \right] \\ f_4(\xi) &= \frac{2\omega_1}{\xi^3 N} \left[(\xi^2 - 1)\left(\frac{2}{A} + \omega_1^2\right) - (\xi - 1)^2 \left(\frac{2}{A} - \omega_1^2\right) \right] \\ N &= \left(\frac{2}{A} + \omega_1^2\right)^2 + \beta^2 \left(\frac{2}{A} - \omega_1^2\right)^2 \end{aligned} \qquad [3.46]$$

The constant A is defined as before in eq. 3.29 as is also the variable T. The other part, S_2, of the solution is obtained by substituting $\tau - \pi/\omega_1$ for τ in the solution (eq. 3.43) for S_1 and remembering that the solution thus obtained for S_2 is valid only for $\tau > \xi - 1 + \pi/\omega_1$.

It is obvious that, in spite of the fact that the solution obtained appears in terms of elementary functions only, the complexity of it makes it necessary to study it by means of a selection of cases which are worked out, and which exhibit some general features.

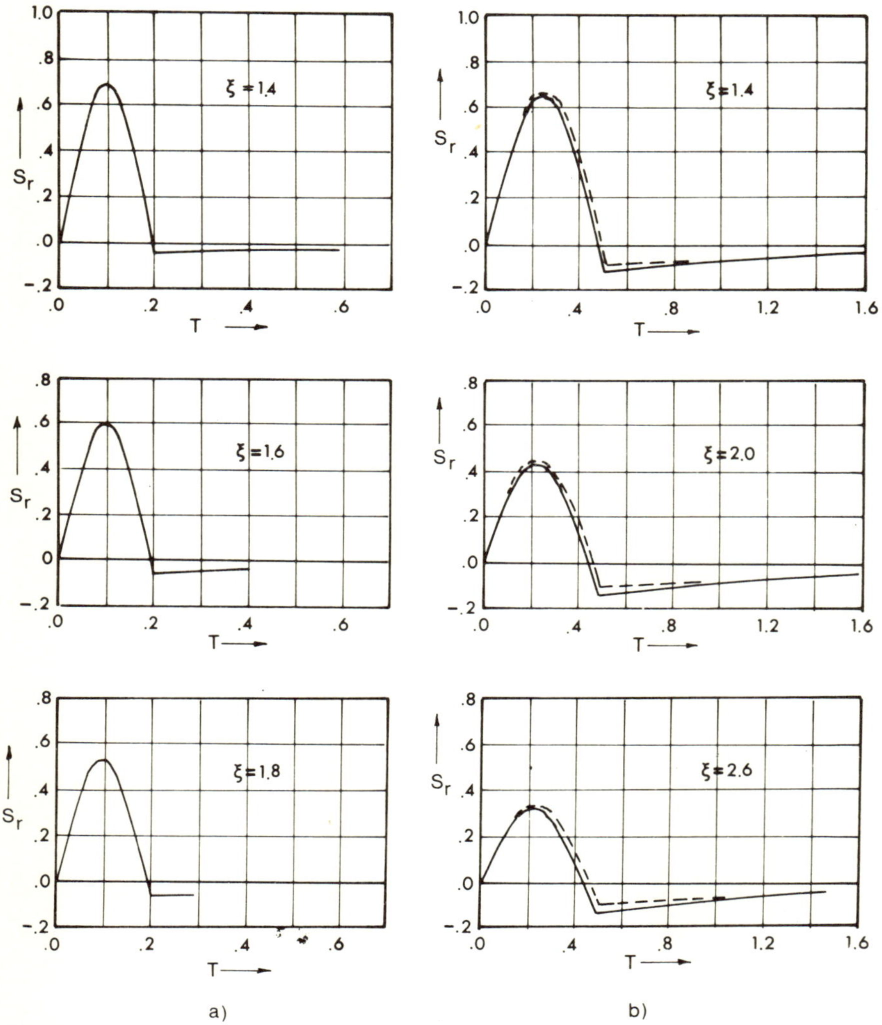

Fig. 19. Pickup signals at different locations (ξ) for a sinusoidal input pulse: (a) τ_0 = 0.2, (b) τ_0 = 0.5.

In Fig. 19a the "pickup signal" (S_r as a function of time τ at a fixed location ξ) is computed for three different locations (ξ = 1.4, 1.6 and 1.8). Because nothing happens at location ξ for $\tau < \xi - 1$, the time is counted in terms of T which is zero when the wave front arrives. (The same is true also for Fig. 20.) The duration of the input pulse is τ_0 = 0.2 and Fig. 19b shows the same for an input pulse duration τ_0 = 0.5. It is seen that in both cases the maximum amplitude of the wave decreases with increasing distance from the cavity. It is furthermore observed that the duration of

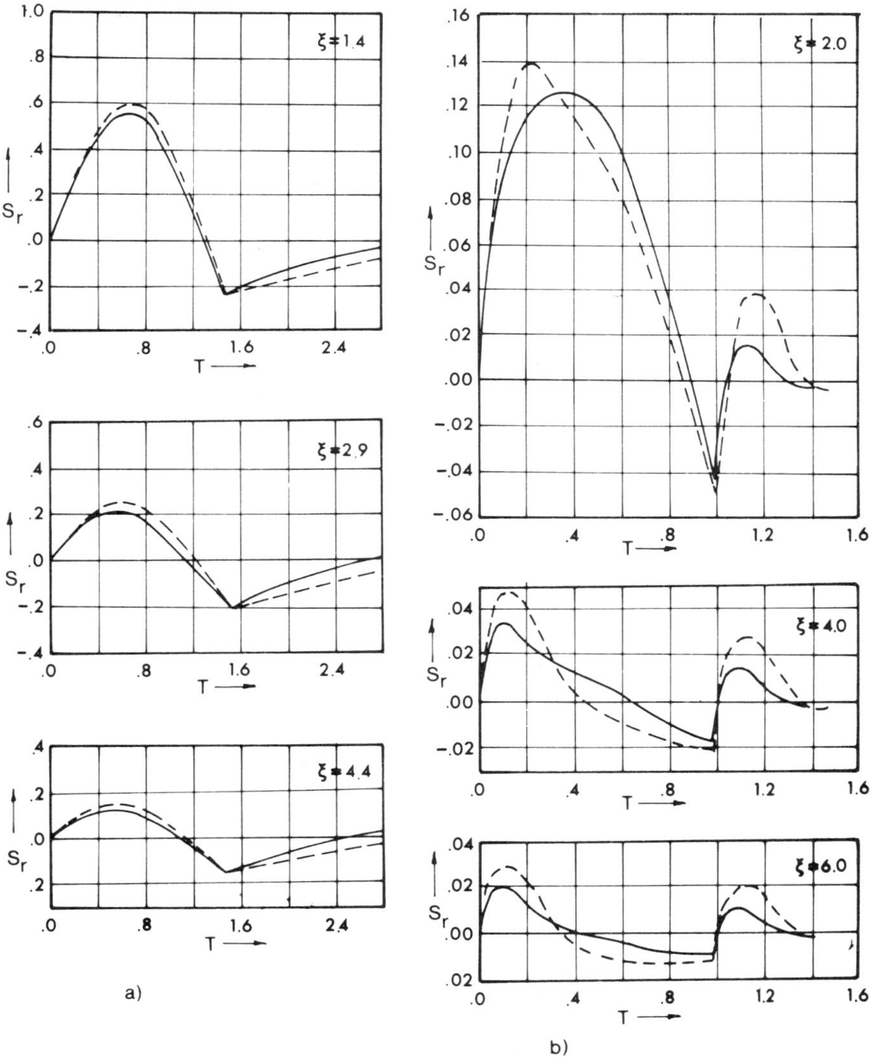

Fig. 20. Pickup signals at different locations (ξ) for a sinusoidal input pulse: (a) τ_0 = 1.5, (b) τ_0 = 10.0.

the positive phase of the wave is smaller than τ_0, and that the difference increases both with increasing distance (ξ) and increasing input duration (τ_0). To show this even more clearly, attention is drawn to Fig. 20, where the duration of the input pulse τ_0 = 1.5 (Fig. 20a) and τ_0 = 10.0 (Fig. 20b). It is seen that the conclusion drawn above is very well exhibited in Fig. 20b where for ξ = 6.0 the duration of the positive phase (the first one) of the wave is $\approx 0.4\ \tau_0$. However, the case of τ_0 = 10 also shows another feature of the problem. For greater values of ξ, the duration of the pulse is so long that the eigenfrequency of the system is allowed time to act, and this, of course, deforms the shape of the pulse. In the case presented here one is seemingly left with two small maximum amplitudes, which are a distance τ_0 apart in time. This reflects the transmission of the two discontinuities of the input pulse, one when the wave starts ($\tau = 0$) and one when the input pulse ends ($\tau = \tau_0$). Thus, it is again shown that discontinuities in the stress situation are very strongly transmitted.

One conclusion may be drawn from this which will have a bearing on later interpretation of experimental results: For not too large values of ξ, the duration of

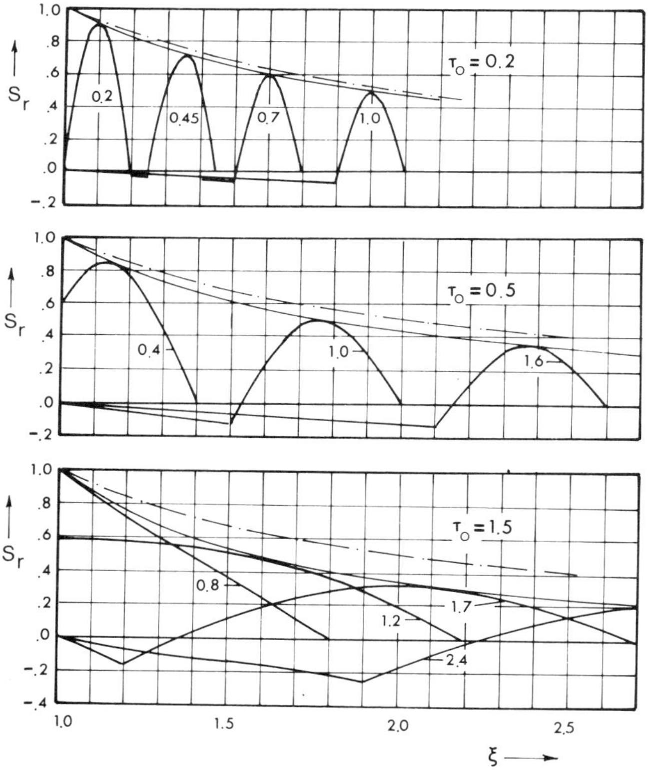

Fig. 21. Stress distribution in the material at different instances (τ) for different values of τ_0.

the positive phase of the pulse measured at various positions ξ may be used as a measure of the duration, τ_0, of the input pulse. Further comments on this point will be given later.

In Fig. 21 the stress situation at different times τ is given by displaying S_r as a function of ξ for three different values of τ_0. The dash-dotted curves represent ξ^{-1}, i.e. they represent the maximum value of the stress (S_r) at each point which would have been created by an impulsively started input pulse (example: rectangular pulse). It is observed that the corresponding curves for the cases in question lie below these curves, which indicates a more rapid attenuation in these cases. The difference increases with increasing values of τ_0. Because of the importance of these attenuation curves, some further comments are appropriate. First it is observed that for the case in question the initial slope of the input pulse in view of 3.35 may be expressed as:

$$\left.\frac{\partial f}{\partial \tau}\right|_{\tau=0} = \omega_1 = \frac{\pi}{\tau_0} \qquad [3.47]$$

τ_0 can here be considered not only as a measure of the duration of the input pulse, but also as a measure of how rapidly the input pulse initially increases ($\tau_0 = 0$ approximates the impulsively started input pulse). Thus, the attenuation is greater for a slowly rising (large τ_0) input pulse than for one which rises rapidly. However, if the situation for larger distances is considered, one obtains a situation as shown in Fig. 22. This shows that even though the initial attenuation of the maximum value of S_r increases with increasing values of τ_0, the situation further away (larger values of ξ) is such that again the maximum value of S_r is inversely proportional to ξ.

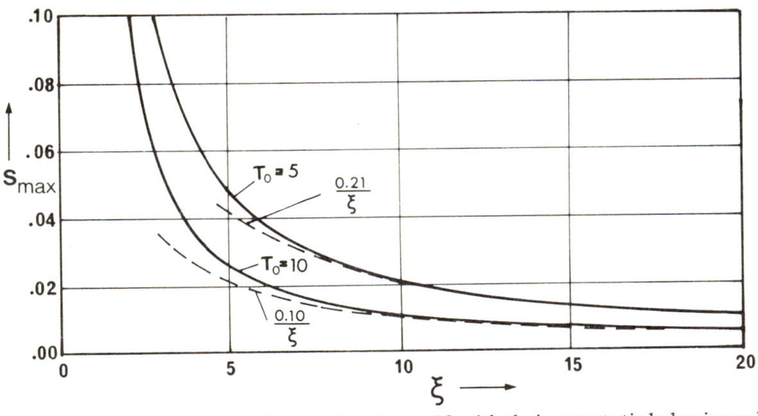

Fig. 22. Attenuation curves for $\tau_0 = 5$ and $\tau_0 = 10$ with their asymtotic behavior exhibited (dotted lines).

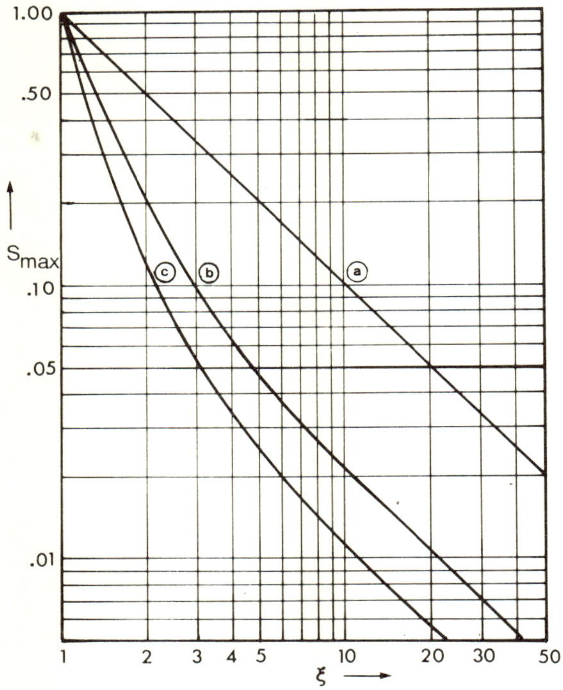

Fig. 23. Attenuation curves for: (a) $\tau_0 = 0$, (b) $\tau_0 = 5.0$, (c) $\tau_0 = 10.0$.

Perhaps the best way to display the entire situation is to exhibit $S_{r,\max}$ at any point created by the input pulse as a function of the distance, as this is done in the double logarithmic diagram in Fig. 23. Here, the curve labeled "a" is valid for all input pulses which are characterized by impulsively attaining their maximum value, irrespective of their later form. Thus, the importance of what may be called the "rising time" of the input pulse is shown. However, with a slowly rising pulse (large τ_0) the attenuation depends on τ_0 as shown by curves "b" and "c" which are computed for $\tau_0 = 5.0$ and $\tau_0 = 10.0$, respectively. The attenuation is, therefore, described not only by the slope of these curves (which, for larger values of ξ, is equal to -1), but also by their vertical position. This position depends on what happens at the initial point $\xi = 1$, i.e. on the input pulse.

Chapter 4

SOME CHARACTERISTIC FEATURES OF A CYLINDRICAL WAVE (ELASTIC MEDIUM)

INTRODUCTION

In this chapter only two examples are examined because of their importance in experimental investigations. The problems discussed are selected also because of the fact that their solutions are relatively simple. Consequently, the problem of a cylindrical wave moving outwards towards infinity from a cylindrical cavity is not discussed because of its mathematical complexity. For this case the reader is referred to the original solution by Selberg (1952).

The main reason for selecting the problems presented here is, that in certain cases pickups in the shape of cylinders are used in the investigations in rocks. It is then imperative that one knows the response of a cylinder to an outside excitation.

BASIC EQUATIONS

The cases examined are all characterized by cylindrical symmetry; the conclusion may be drawn that no quantity will exhibit any φ-dependence, nor any z-dependence in the cylindrical coordinates (r, φ, z). Furthermore, one may also introduce the condition on the displacement field, that components v_φ and v_z of the displacement vector $\vec{s}$ are equal to zero. Then, with only component v_r different from zero, one will have (from 1.46, 1.47 and 1.48):

Strain:

$$\epsilon_r = \frac{\partial v_r}{\partial r}, \qquad \epsilon_\varphi = \frac{v_r}{r}, \qquad \epsilon_z = 0 \qquad [4.1]$$

Shear strain:

$$\gamma_{r\varphi} = \gamma_{\varphi z} = \gamma_{zr} = 0 \qquad [4.2]$$

Equation of motion:

$$\frac{\partial \sigma_r}{\partial r} + \frac{\sigma_r - \sigma_\varphi}{r} = \rho \frac{\partial^2 v_r}{\partial t^2} \qquad [4.3]$$

The relation between the state of stress and the state of deformation then renders:

$$I_1 = \sigma_\varphi + \sigma_\varphi + \sigma_z = (2G + 3\lambda)\ \mathrm{div}\,\vec{s} = (2G + 3\lambda)\left(\frac{\partial v_r}{\partial r} + \frac{v_r}{r}\right) \qquad [4.4]$$

and:

$$\left.\begin{aligned} \sigma_r - \tfrac{1}{3}I_1 &= 2G\left[\frac{\partial v_r}{\partial r} - \frac{1}{3}\left(\frac{\partial v_r}{\partial r} + \frac{v_r}{r}\right)\right] \\ \sigma_\varphi - \tfrac{1}{3}I_1 &= 2G\left[\frac{v_r}{r} - \frac{1}{3}\left(\frac{\partial v_r}{\partial r} + \frac{v_r}{r}\right)\right] \\ \sigma_z - \tfrac{1}{3}I_1 &= -\frac{2G}{3}\left(\frac{\partial v_r}{\partial r} + \frac{v_r}{r}\right) \end{aligned}\right| \qquad [4.5]$$

Explicitly the stresses will be:

$$\tau_{r\varphi} = \tau_{\varphi z} = \tau_{zr} = 0 \qquad [4.6]$$

and:

$$\left.\begin{aligned} \sigma_r &= 2G\frac{\partial v_r}{\partial r} + \lambda\left(\frac{\partial v_r}{\partial r} + \frac{v_r}{r}\right) \\ \sigma_\varphi &= 2G\frac{v_r}{r} + \lambda\left(\frac{\partial v_r}{\partial r} + \frac{v_r}{r}\right) \\ \sigma_z &= \lambda\left(\frac{\partial v_r}{\partial r} + \frac{v_r}{r}\right) \end{aligned}\right| \qquad [4.7]$$

Introduction of 4.7 into 4.3 will then finally give the differential equation for v_r:

$$(2G + \lambda)\frac{\partial}{\partial r}\left[\frac{\partial v_r}{\partial r} + \frac{v_r}{r}\right] = \rho\,\frac{\partial^2 v_r}{\partial t^2} \qquad [4.8]$$

It is seen that also in this case the velocity, c, of the compression wave enters the equation as a parameter.

Now, under the assumptions made, the situation will always be such that an infinitely extended cylindrical surface will exist, on which a specified stress p will act. Again this stress is assumed to be given by 3.1. If the radius of the cylindrical surface is a, the following non-dimensional quantities may be introduced:

Non-dimensional distance: $\xi = \dfrac{r}{a}$

Non-dimensional time: $\tau = \dfrac{ct}{a}$

Non-dimensional displacement: $\Phi = \dfrac{2G}{p_0} \cdot \dfrac{v_r}{a}$

Non-dimensional stress: $S_r = \dfrac{\sigma_r}{p_0}$

[4.9]

Then, the following expression for S_r is obtained from 4.7:

$$S_r = \frac{\partial \Phi}{\partial \xi} + \frac{\nu}{1 - 2\nu}\left(\frac{\partial \Phi}{\partial \xi} + \frac{\Phi}{\xi}\right) \qquad [4.10]$$

and the differential equation [4.8] converts into:

$$\frac{\partial}{\partial \xi}\left[\frac{\partial \Phi}{\partial \xi} + \frac{\Phi}{\xi}\right] = \frac{\partial^2 \Phi}{\partial \tau^2} \qquad [4.11]$$

This equation may now be solved for the following initial conditions:

$$\Phi(\xi, 0) = 0, \quad \left.\frac{\partial \Phi}{\partial \tau}\right|_{\tau=0} = 0 \qquad [4.12]$$

In view of these conditions the application of the Laplace transform to 4.11 will give:

$$\frac{\partial^2 \Phi^*}{\partial \xi^2} + \frac{1}{\xi}\frac{\partial \Phi^*}{\partial \xi} - \left(\frac{1}{\xi^2} + s^2\right)\Phi^* = 0 \qquad [4.13]$$

The solution to this equation may be expressed in terms of the modified Bessel functions as follows:

$$\Phi^*(\xi, s) = C_1 I_1(s\xi) + C_2 K_1(s\xi) \qquad [4.14]$$

The Laplace transform applied to 4.10 will give:

$$S_r^*(\xi, s) = \frac{\partial \Phi^*}{\partial \xi} + \frac{\nu}{1 - 2\nu}\left(\frac{\partial \Phi^*}{\partial \xi} + \frac{\Phi^*}{\xi}\right) \qquad [4.15]$$

If expression 4.14 is introduced into 4.15 and the standard relationships between Bessel functions and their derivatives are considered, the following solution for S_r^*

is finally obtained:

$$S_r^* = C_1 As\left[I_0(s\xi) - \frac{1}{As\xi}\,I_1(s\xi)\right] - C_2 As\left[K_0(s\xi) + \frac{1}{As\xi}\,K_1(s\xi)\right] \qquad [4.16]$$

where A is defined in eq. 3.29. This is as far as the deductions can be carried unless the boundary conditions are specified, and these will be different for the different cases.

THE INWARD-MOVING PULSE (SOLID CYLINDER)

The case under consideration here will be the one in which at time $\tau = 0$ an evenly distributed pressure is suddenly applied to the outside surface of a circular cylinders with a radius a as shown in Fig. 24. The actual development of p with time is given, as stated earlier, by $f(t)$ in 3.1 and for the case under consideration it is assumed to be a Heaviside function. The cylinder is solid; the boundary conditions may therefore be specified in terms of $\Phi(\xi,\tau)$ and $S_r(\xi,\tau)$ as:

$$\left.\begin{aligned} &\Phi(0,\tau) = 0 \\ &S_r(1,\tau) = f\left(\frac{ct}{a}\right) = \begin{cases} 1 & \tau > 0 \\ 0 & \tau < 0 \end{cases} \end{aligned}\right| \qquad [4.17]$$

The Laplace transform of these conditions will be:

$$\left.\begin{aligned} &\Phi^*(0,s) = 0 \\ &S_r^*(1,s) = f^* = \frac{1}{s} \end{aligned}\right| \qquad [4.18]$$

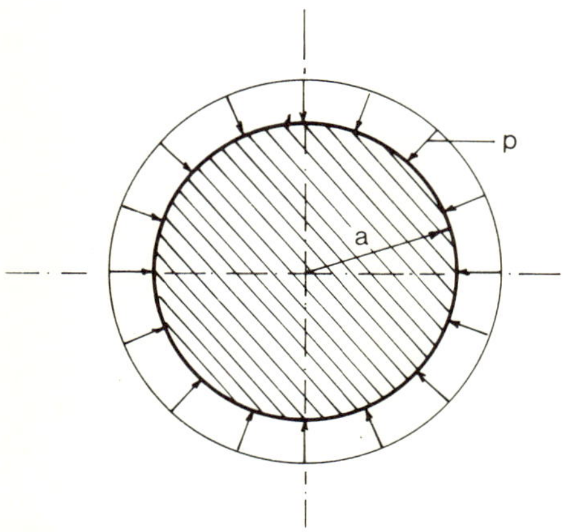

Fig. 24. Cylinder with outside pressure.

Eq. 4.14 is considered first and the behavior of this solution for small values of ξ is given by the series expansions for the modified Bessel functions:

$$I_n(z) = \frac{(\frac{1}{2}z)^n}{(n+1)}\left[1 + \frac{(\frac{1}{2}z)^2}{n+1} + \ldots\right]$$

$$K_n(z) = \tfrac{1}{2}(\tfrac{1}{2}z)^{-n} \sum_{r=0}^{n-1} \frac{(-1)^r (n-r-1)}{r} (\tfrac{1}{2}z)^{2r} + \qquad [4.19]$$

$$+ (-1)^{n+1} (\tfrac{1}{2}z)^n \sum \frac{(\frac{1}{2}z)^{-2r}}{r(n+r)} \{\log(\tfrac{1}{2}z) - \tfrac{1}{2}[\Psi(r+1) - \Psi(m+r+1)]\}$$

where:

$$\Psi(r+1) = \left(1 + \frac{1}{2} + \frac{1}{3} + \ldots + \frac{1}{r}\right) - \gamma \quad (\gamma = 0.5772) \qquad [4.20]$$

From these expressions it is evident that:

$$\lim_{\xi \to 0} I_1(s\xi) = 0$$

$$\lim_{\xi \to 0} K_1(s\xi) = \infty \qquad [4.21]$$

Thus, it may be concluded that $C_2 = 0$ in eq. 4.14 if 4.18 is to be satisfied. It is a matter of routine algebra to show that the second condition in 4.18 will be satisfied if C_1 is chosen such that:

$$S_r^*(\xi, s) = f^*(s) \frac{I_0(s\xi) - \frac{1}{As\xi} I_1(s\xi)}{I_0(s) - \frac{1}{As} I_1(s)} \qquad [4.22]$$

This expression is valid for any time dependence $f(t)$ of the input pulse but here $f^*(s) = 1/s$ will be used in accordance with 4.18. The state of stress as defined by S_r is finally found from the inverse Laplace transform:

$$S_r(\xi, \tau) = \frac{1}{2\pi i} \int_{b+i\infty}^{b-i\infty} \frac{1}{s} \cdot \frac{I_0(s\xi) - \frac{1}{As\xi} I_1(s\xi)}{I_0(s) - \frac{1}{As} I_1(s)} e^{+s\tau} \, ds \qquad [4.23]$$

The integral in this expression may be evaluated by summing up the residues at the poles of the integrand. These poles will be located at the points $s = s_n$ where s_n are the roots of the equation:

$$I_0(s_n) - \frac{1}{As_n} I_1(s_n) = 0 \qquad [4.24]$$

It may be shown that these roots are located on the imaginary axis in the s-plane, and by introducing:

$$s_n = iy_n \qquad [4.25]$$

eq. 4.24 will be expressible in terms of the ordinary Bessel functions of the first kind:

$$J_0(y_n) - \frac{1}{Ay_n} J_1(y_n) = 0 \qquad [4.26]$$

The first two roots, y_1 and y_2, of this equation must be found by iteration, but for $n > 2$ one may use the approximate value:

$$y_n = \frac{\pi}{8}(4n + 3)\left[1 + \sqrt{1 - \frac{8(8 - A)}{A\pi^2(4n + 3)^2}}\right] \qquad [4.27]$$

The final expression for S_r is then found as a Fourier series as follows:

$$S_r(\xi, \tau) = 1 - 2\sum_{n=0}^{\infty} \frac{\left[J_0(\xi y_n) - \dfrac{1}{A\xi y_n} J_1(\xi y_n)\right] A}{J_0(y_n) + \left(Ay_n - \dfrac{2}{y_n}\right) J_1(y_n)} \cos(y_n \tau) \qquad [4.28]$$

One comment should be made here. The solution is given in the form of a Fourier series, and such series give a very slow convergence in the neighborhood of discontinuities. Because this solution must exhibit a discontinuity at the front of the wave, one cannot expect the values close to the front to be accurate. The series has, therefore, been evaluated with 300 terms in the results which will be exhibited subsequently.

In Fig. 25 stress S_r is shown as a function of ξ at regular intervals of τ. It is seen how the value of S_r at the shock front steadily increases as it approaches the center of the cylinder, and actually becomes infinite at $\xi = 0$ (at time $\tau = 1$). There is thus

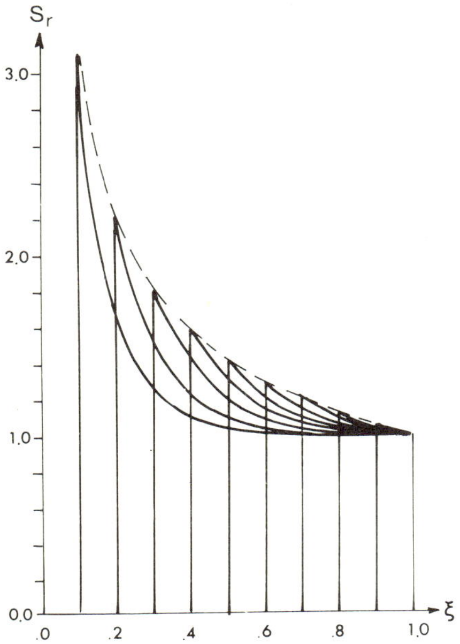

Fig. 25. Stress distribution in the cylinder at regular intervals of τ ($\tau < 1$).

a "focusing" effect for the stress wave which is due to the geometry. In the case of a shock wave moving outwards from a spherical cavity the so-called "geometric attenuation" was observed. Here a "negative geometric attenuation" is observed.

In spite of the fact that the solution passes through a singularity at $\xi = 0$, the situation for $\tau > 1$ may be obtained from the solution. In order to show the complexity of the solution, it is displayed in a series of consecutive instances in Fig. 26a–i. The process is very similar to what happens in a slab (Fig. 8), and (a) and (b) show the inward-moving shock fronts with increasing maximum values as they get closer to the center. In (c) and (d) it is seen that, when the two fronts have passed each other, they leave a region behind in which $S_r \approx 2$ just as in the case of the slab. In (e) the reflection process starts, and it is noticed that there is a difference between the stress situation here and that which exists in the case of the slab. In (f) and (g) the reflected waves move inward and the stress situation at the center is characterized by a gradual decrease towards zero as shown in (i).

THE NORMAL REFLECTION OF A CYLINDRICAL WAVE

The reflection of a cylindrical wave moving normal to the free cylindrical surface at which the reflection takes place may be studied by means of the same solution as the one described in the section *Basic equations* of this chapter. The only change will be in the boundary conditions, which may be adopted from Fig. 27. At the outer surface of a hollow cylinder ($\xi = 1$) a pressure is suddenly applied, given by:

$$p = p_0 f(\tau) \qquad [4.29]$$

At the inner surface ($\xi = \xi_0$) the normal stress is zero. Thus, the ordinary procedure will lead to the following boundary conditions for $S_r^*(\xi,s)$:

$$S_r^*(\xi_0, s) = 0$$

$$S_r^*(1, s) = f^*(s) \qquad [4.30]$$

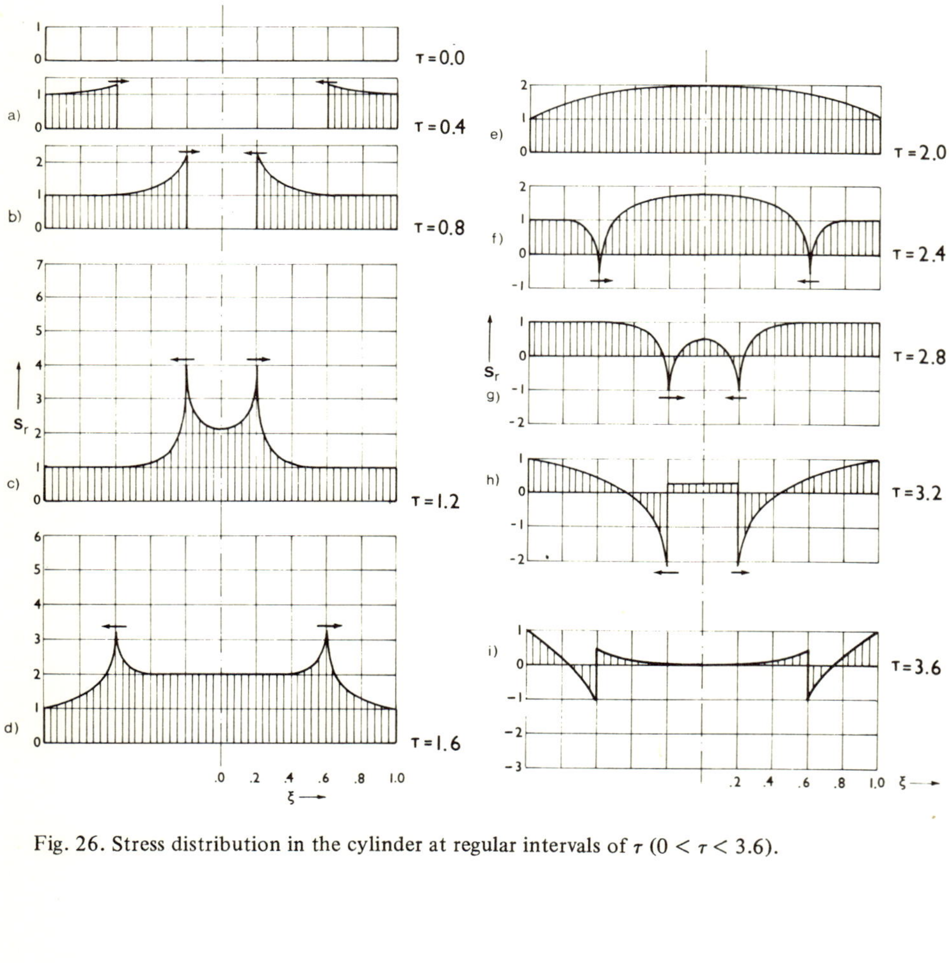

Fig. 26. Stress distribution in the cylinder at regular intervals of τ ($0 < \tau < 3.6$).

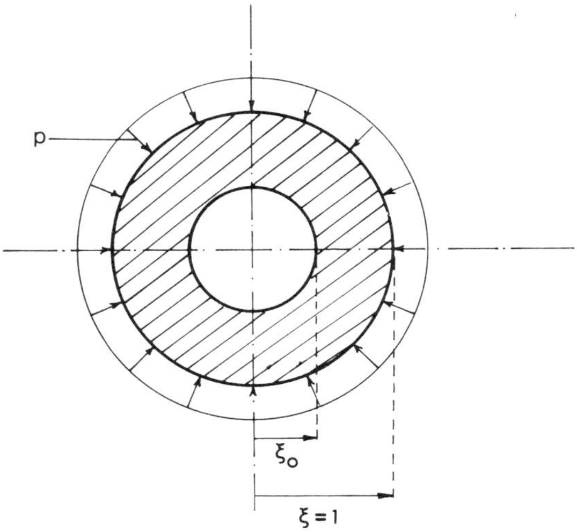

Fig. 27. Outside pressure applied to a hollow cylinder.

These two conditions are sufficient to determine the constants of integration in 4.16 and the solution for S_r^* is complete. In addition the shape of the input pulse (given by $f(\tau)$) must be specified in order to find $f^*(s)$. This whole procedure and the final evaluation of the inverse Laplace transform are straightforward but tedious and have been omitted here. The reflection process is better exhibited by worked out cases, and below two cases will be shown. In both cases it should, however, be remembered that the reflection must now be exhibited not only for stress S_r and particle velocity $\partial\Phi/\partial\tau$ but also for the hoop stress, S_φ.

Case 1

The first case is the one in which a rectangular pulse with a duration τ_0 is suddenly applied to the outer surface ($\xi = 1$). The function $f(\tau)$ will then be specified as:

$$f(\tau) = \left\{\begin{matrix} 0, \tau < 0 \\ 1, 0 < \tau < \tau_0 \\ 0, \tau > \tau_0 \end{matrix}\right\} \rightarrow f^*(s) = \frac{1}{s}(1 - e^{-s\tau_0}) \qquad [4.31]$$

Case 2

In the second case the input pulse is defined as a triangular pulse as shown in

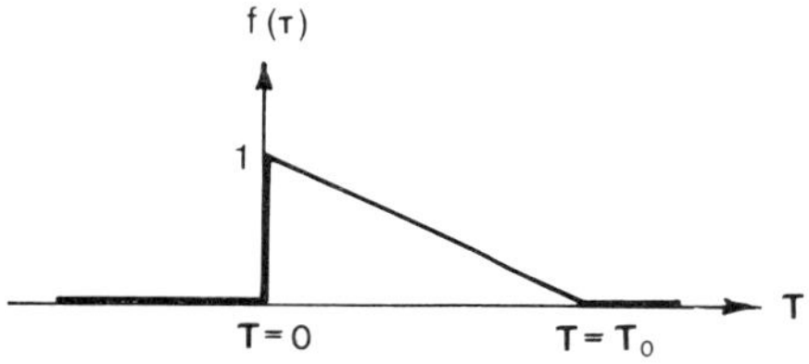

Fig. 28. The triangular input pulse.

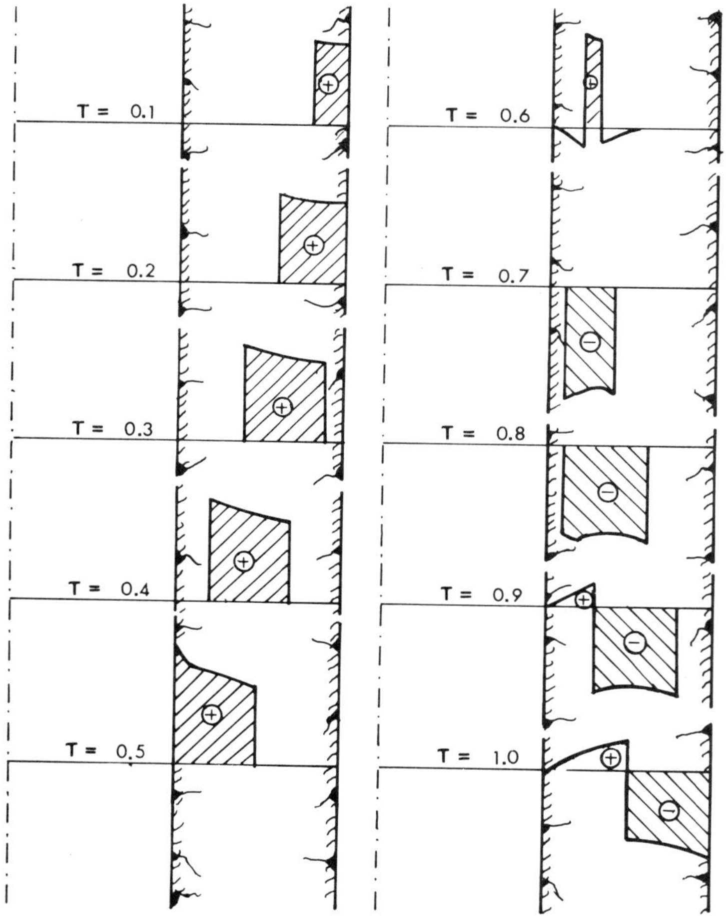

Fig. 29. Distribution of the radial stress, S_r, at consecutive instances, τ, in a hollow cylinder excited by a rectangular input pulse. $\xi_0 = 0.5$, $\tau_0 = 0.25$, $\nu = 0.0$.

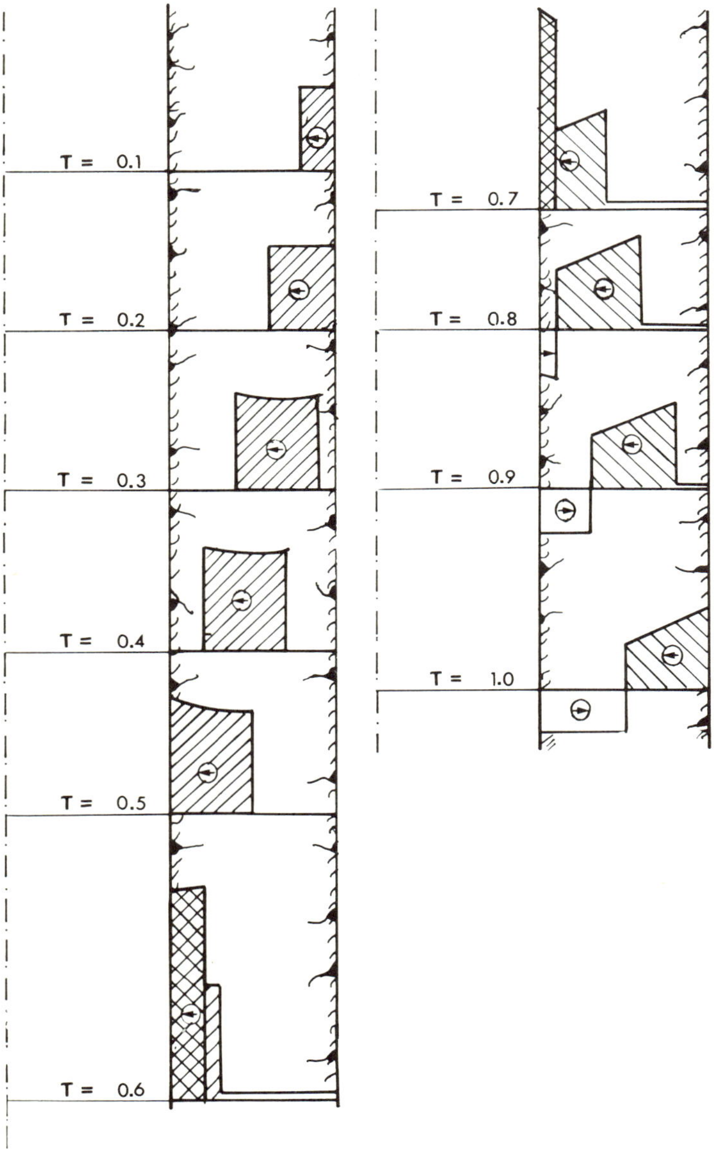

Fig. 30. Distribution of the radial particle velocity at consecutive instances, τ, in a hollow cylinder excited by a rectangular input pulse. $\xi_0 = 0.5$, $\tau_0 = 0.25$, $\nu = 0.0$.

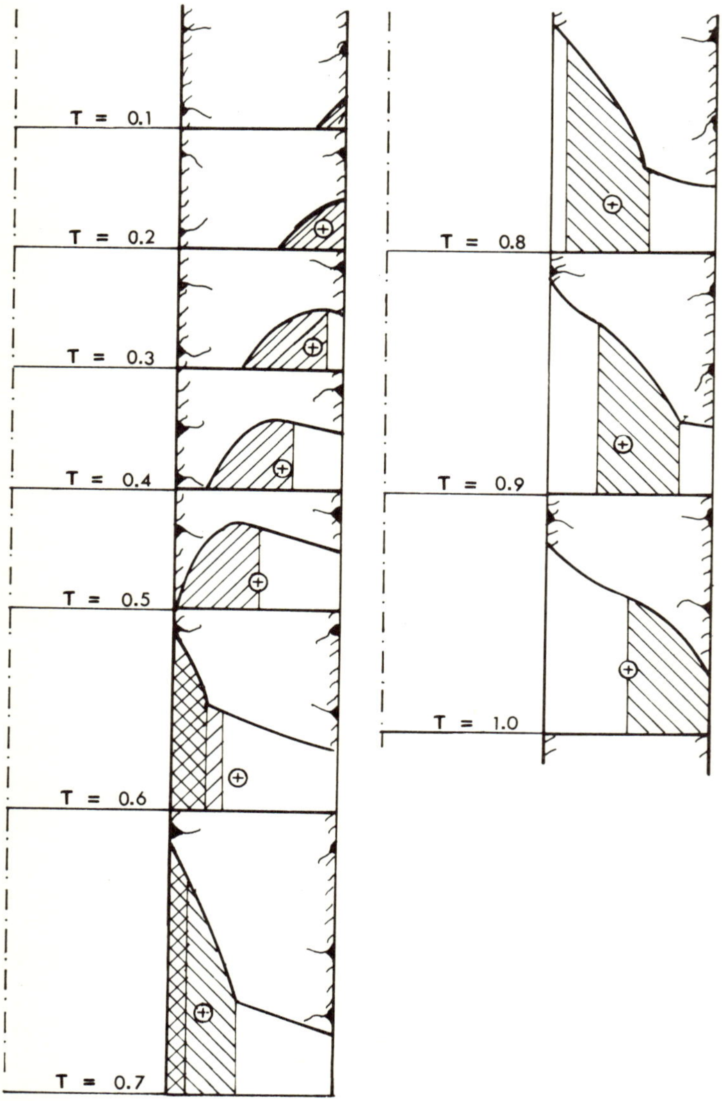

Fig. 31. Distribution of the hoop stress, S_φ, at consecutive instances, τ, in a hollow cylinder excited by a rectangular input pulse. $\xi_0 = 0.5$, $\tau_0 = 0.25$, $\nu = 0.0$.

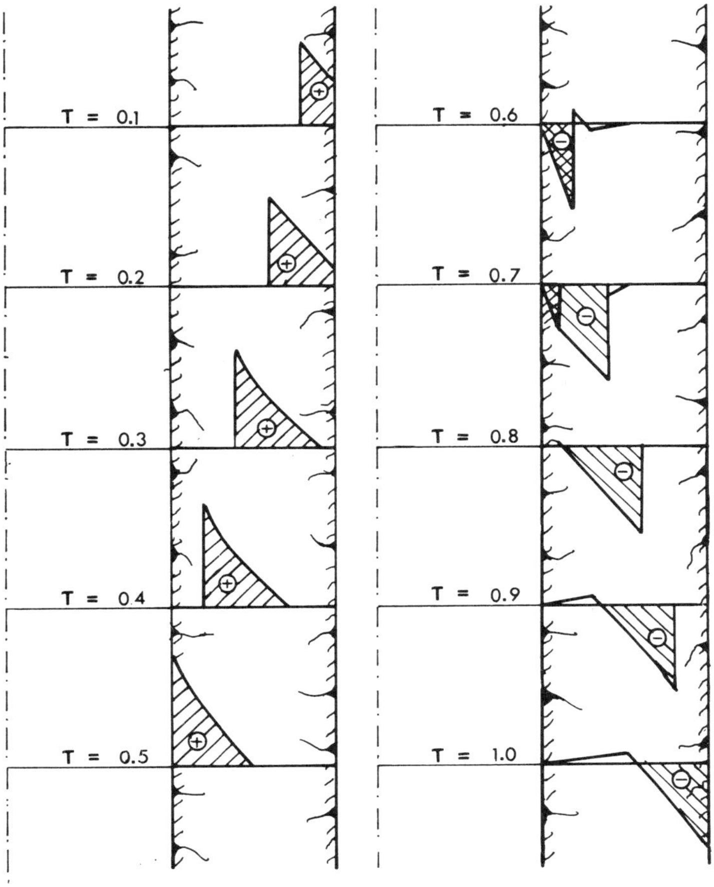

Fig. 32. Distribution of the radial stress, S_r, at consecutive instances, τ, in a hollow cylinder excited by a triangular input pulse. $\xi_0 = 0.5$, $\tau_0 = 0.25$, $\nu = 0.0$.

Fig. 28. The function $f(\tau)$ is then defined as:

$$f(\tau) = \begin{cases} 0 & \tau < 0 \\ 1 - \dfrac{\tau}{\tau_0}, & 0 < \tau < \tau_0 \\ 0 & \tau > \tau_0 \end{cases} \qquad [4.32]$$

which gives:

$$f^*(s) = \left(\frac{1}{s} - \frac{1}{\tau_0 s^2}\right) + \frac{1}{\tau_0 s^2}\, e^{-s\tau_0} \qquad [4.33]$$

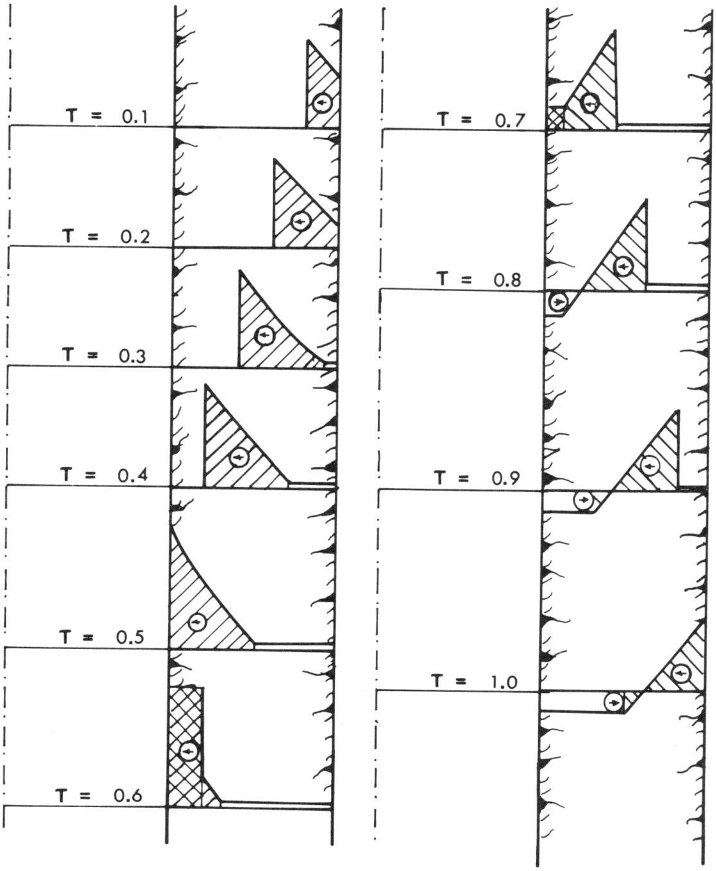

Fig. 33. Distribution of the particle velocity at consecutive instances, τ, in a hollow cylinder excited by a triangular input pulse. $\xi_0 = 0.5$, $\tau_0 = 0.25$, $\nu = 0.0$.

In both cases $\tau_0 = 0.25$ has been chosen for the numerical examples which are exhibited in Figs. 29–34. It is hoped that these are self explanatory, but a few remarks may be appropriate. The reflection procedure for the radial stress, S_r, is seen to analogously follow the cases described in Chapter 2, Figs. 9 and 12. The same holds true for the particle velocity, compare Chapter 2, Fig. 11. The hoop stress, S_φ, however, is seen not to reflect in the same way as the radial stress. That this is correct may be concluded from 4.7. Assuming Poisson's ratio $\nu = 0$ one sees that $\sigma_\varphi \sim v_r$. At the inner surface the particle velocities due to the compression wave and due to the "mirror" image tension wave are of the same sign and thus the two are added as shown for the plane case. Here the same takes place, but that means that also v_r from the two waves will have to be added. Thus, the hoop

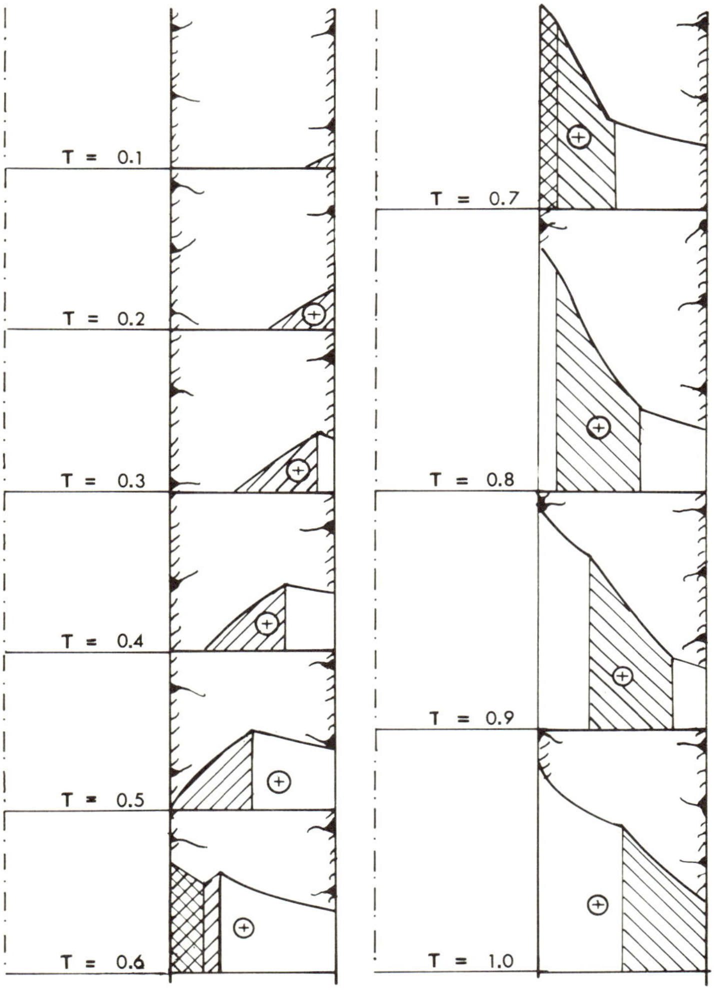

Fig. 34. Distribution of the hoop stress, S_φ, at consecutive instances, τ, in a hollow cylinder excited by triangular input pulse. $\xi_0 = 0.5$, $\tau_0 = 0.25$, $\nu = 0.0$.

stress, S_φ, reflects like the particle velocity as shown in Figs. 31 and 34.

A final remark may be made on the spalling phenomenon. If it is assumed that the material of the hollow cylinder can sustain large stresses in compression but breaks at a non-dimensional stress $S = 0.25$, then in Fig. 29 a cylindrical shell of a thickness $0.1a$ is chipped off at time $\tau = 0.6$. Fig. 30 shows that at that time the shell has a high velocity directed inward. Thus, in the case of rock material, the shell would break and be thrusted inwards into the cavity.

Chapter 5

THE HARMONIC INPUT PULSE (ELASTIC MEDIUM)

INTRODUCTION

In the preceding chapters the emphasis has been placed on cases in which the outside excitation has consisted of a pulse of finite duration. This has been done mainly to exhibit cases which bear a resemblance to the situation in rock materials in which shock waves usually are created by detonating charges. However, the case of an outside excitation in the form of a harmonically oscillating stress has so many features in it of great physical significance, that this case is examined separately in the present chapter. Again, physical realities will be brought out by chosen examples.

THE PLANE CASE

Reference is here made to Chapter 2, section *The plane wave in a finite medium,* where an infinite slab with a thickness of $2h$ is shown in Fig. 7. Instead of the former input pulse it will here be assumed that a stress σ, which oscillates with time, is applied on the surfaces of the slab:

$$\sigma = \sigma_0 \sin(\omega t) \qquad [5.1]$$

The basic equations remain the same as before, and keeping the non-dimensional quantities in 2.36 unchanged, one is left with the following mathematical formulation of the problem.

Differential equation:

$$\frac{\partial^2 S_x}{\partial \xi^2} = \frac{\partial^2 S_x}{\partial \tau^2} \qquad [5.2]$$

Boundary condition for $S_x(\xi, \tau)$:

$$\left. \begin{aligned} S_x(+1, \tau) &= \sin(\Omega\tau) \\ S_x(-1, \tau) &= \sin(\Omega\tau) \end{aligned} \right| \qquad [5.3]$$

where:

$$\Omega = \frac{\omega h}{c} \qquad [5.4]$$

In addition, the initial conditions ought to be specified. However, these will here be replaced by the assumption that S_x oscillates with the same frequency as the imposed frequency, i.e. a solution is sought such that:

$$S_x = F(\xi) \cdot \sin(\Omega\tau) \qquad [5.5]$$

The physical reality behind this is the assumption that a "quasi-static" state will be reached irrespective of the way in which the oscillation was initiated.

By introduction of eq. 5.5 into 5.2 one is left with the following differential equation for F:

$$\frac{d^2F}{d\xi^2} = -\Omega^2 F \qquad [5.6]$$

with the boundary conditions:

$$\left.\begin{aligned} F(+1) &= 1 \\ F(-1) &= 1 \end{aligned}\right| \qquad [5.7]$$

whereupon the final solution to the problem becomes:

$$S_x = \frac{\cos(\Omega\xi)}{\cos(\Omega)} \sin(\Omega\tau) \qquad [5.8]$$

This solution reveals that $S_x \to \infty$ for:

$$\Omega = \Omega_n = \frac{2n+1}{2}\pi \qquad (n = 0.1, 2, \ldots) \qquad [5.9]$$

In view of eq. 5.4 this means that the system (slab) has an infinite number of eigenfrequencies $\omega_n = (2n+1)/2 \cdot \pi c/h$ and the stress $S_x \to \infty$ when the imposed frequency $\omega \to \omega_n$.

It should be noticed that the lowest eigenfrequency (ω_0) corresponds to the frequency found for the slab in Chapter 2, section *Plane waves in a finite medium.*

The solution is plotted in Fig. 35 for five instances within the first quarter period with $\Omega < \Omega_0$ and $\Omega_0 < \Omega < \Omega_1$. This illustrates how so-called "standing" stress waves are created, and the slab is divided by "nodal planes" where the stress is always zero (dashed lines in Fig. 35). For $\Omega_n < \Omega < \Omega_{n+1}$ the number of nodal planes will be $2n$.

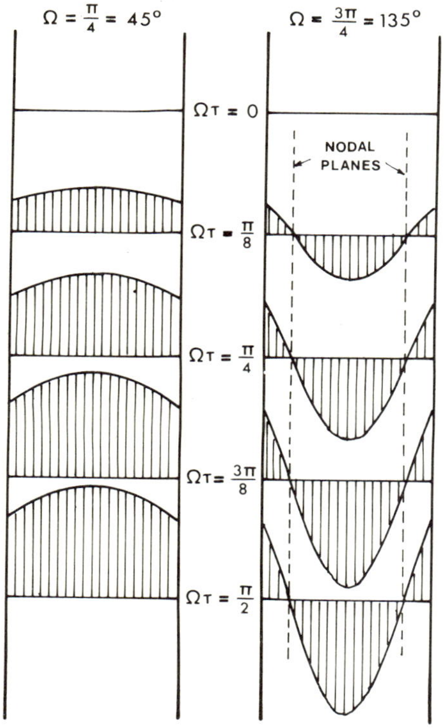

Fig. 35. Stress distribution in a slab with an oscillating stress applied to the slabs surfaces. Imposed frequencies: $\Omega = \pi/4$, $\Omega = 3\pi/4$.

THE CYLINDRICAL CASE

In this case an infinite circular cylinder is considered on the surface of which an evenly distributed pressure, p, acts, which is a harmonic function of time. The situation is identical to the case considered in Chapter 4, section *The inward-moving pulse (solid cylinder)*, with the exception that p is now given as:

$$p = p_0 \sin(\omega t) \qquad [5.10]$$

The differential equation for the non-dimensional displacement Φ (defined in 4.9) will be identical to 4.11:

$$\frac{\partial}{\partial \xi}\left(\frac{\partial \Phi}{\partial \xi} + \frac{\Phi}{\xi}\right) = \frac{\partial^2 \Phi}{\partial \tau^2} \qquad [5.11]$$

with the boundary conditions:

$$\Phi(0, \tau) = 0$$

$$S_r(1, \tau) = \left[\frac{\partial \Phi}{\partial \xi} + \frac{\nu}{1-2\nu}\left(\frac{\partial \Phi}{\partial \xi} + \frac{\Phi}{\xi}\right)\right]_{\xi=1} = \sin(\Omega\tau) \quad [5.12]$$

where:

$$\Omega = \frac{a\omega}{c} \quad [5.13]$$

In addition to these conditions, initial conditions ought to be specified. However, as in the previous section these will be replaced by the assumption that S_r and Φ oscillate with the same frequency as the imposed frequency, i.e. a solution is sought such that:

$$S_r = G(\xi) \cdot \sin(\Omega\tau)$$

$$\Phi = F(\xi) \cdot \sin(\Omega\tau) \quad [5.14]$$

Again the physical reality behind this assumption is the same as before.

In view of 4.10, the following relation will exist between the two functions G and F:

$$G(\xi) = \frac{dF}{d\xi} + \frac{\nu}{1-2\nu}\left(\frac{dF}{d\xi} + \frac{F}{\xi}\right) \quad [5.15]$$

and the differential equation [5.11] will give:

$$\frac{d^2F}{d\xi^2} + \frac{1}{3}\frac{dF}{d\xi} - \left(\frac{1}{\xi^2} - \Omega^2\right) F = 0 \quad [5.16]$$

The boundary conditions are obtained from 5.12:

$$F(0) = 0$$

$$G(1) = \left[\frac{dF}{d\xi} + \frac{\nu}{1-2\nu}\left(\frac{dF}{d\xi} + \frac{F}{\xi}\right)\right]_{\xi=1} = 1 \quad [5.17]$$

A comparison between the formulation of this problem and the differential equation [4.13] with its boundary conditions [4.18] shows that the two problems are identical if the following substitutions are made:

$$f^*(s) \Rightarrow 1$$

$$s \Rightarrow i\Omega \quad [5.18]$$

The solution to the present problem is thus immediately obtained from 4.22 by introduction of 5.18:

$$S_r = G(\xi)\sin(\Omega\tau) = \frac{J_0(\Omega\xi) - \frac{1}{A\Omega\xi}J_1(\Omega\xi)}{J_0(\Omega) - \frac{1}{A\Omega}J_1(\Omega)}\sin(\Omega\tau) \qquad [5.19]$$

This solution exhibits the same phenomenon as the previous one, inasmuch as $S_r \to \infty$ when:

$$\Omega \to \Omega_n = y_n \qquad [5.20]$$

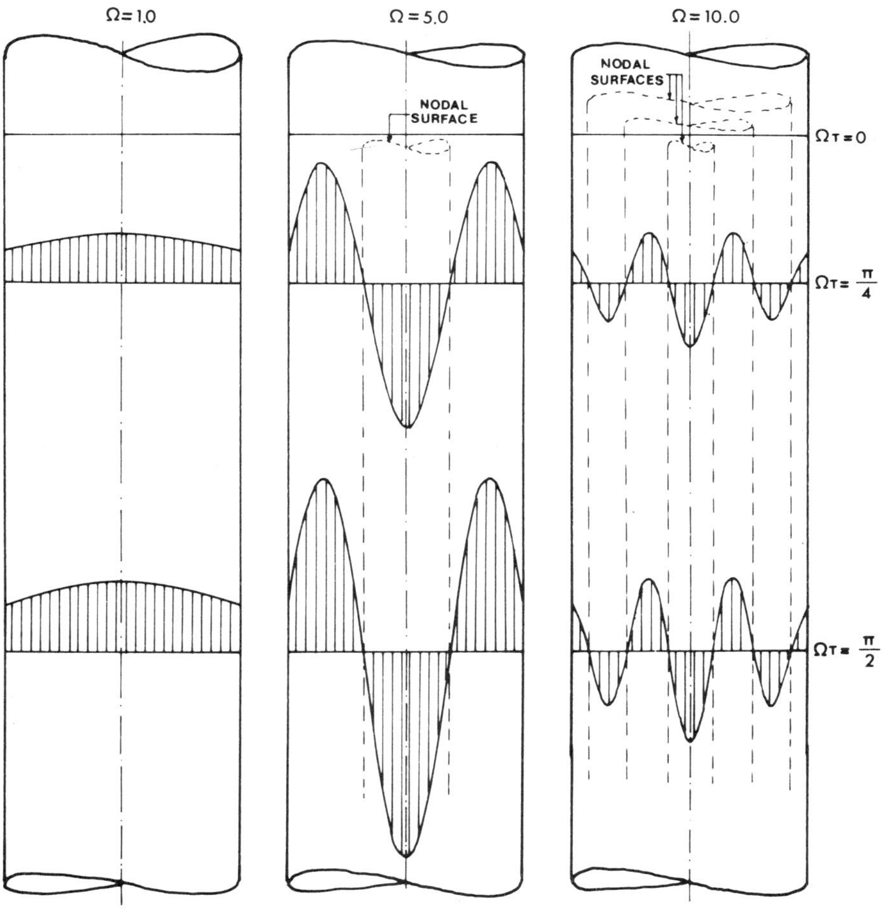

Fig. 36. Standing stress waves in a cylinder created by harmonically oscillating surface traction ($A = 1$).

where y_n is defined as the roots to eq. 4.26. These roots thus represent the eigen-frequencies of the system (cylinder).

The solution 5.19 is exhibited in Fig. 36 for chosen values of the input frequency Ω. It is seen how standing stress waves are created in the cylinder just as the case was in the slab in Fig. 35. In this case cylindrical "nodal" surfaces may occur as shown.

THE SPHERICAL CASE

The present case will be different from the two preceding ones in as much as the waves shall be transmitted outwards in an infinite medium, and not inwards in a finite medium as before. The situation to be studied is identical to the problem in Chapter 3 with p in 3.1 now being specified as:

$$p = p_0 \sin(\omega t) \qquad [5.21]$$

This means that the differential equation for the non-dimensional displacement Φ will be as given in 3.10:

$$\frac{\partial^2 \Phi}{\partial \xi^2} + \frac{2}{\xi}\frac{\partial \Phi}{\partial \xi} - \frac{2}{\xi^2}\Phi = \frac{\partial^2 \Phi}{\partial \tau^2} \qquad [5.22]$$

with the boundary conditions:

$$\left.\begin{aligned} S_r(1, \tau) &= \sin(\Omega\tau) \\ S_r(\infty, \tau) &= 0 \end{aligned}\right| \qquad [5.23]$$

where:

$$\Omega = \frac{a\omega}{c} \qquad [5.24]$$

If one here starts out to replace the initial conditions with the assumption that one may start looking directly for the semi-steady case (as was done in the two previous sections) one will find that the solution remains indeterminate. Thus in the present case the final semi-steady case is determined by the way in which the excitation starts.

The initial conditions to be adopted here will be the same as those given in Table 1 on p. 37. The input pulse is now:

$$f\left(\frac{ct}{a}\right) = \sin(\Omega\tau) \quad \text{for } \tau > 0 \qquad [5.25]$$

which gives:

$$f^*(s) = \frac{\Omega}{s^2 + \Omega^2} \qquad [5.26]$$

Thus the problem is exactly the same as that solved in Chapter 3, section *The sinusoidal input pulse;* the only difference being that the solution now consists only of S_1 in 3.39 or 3.43.

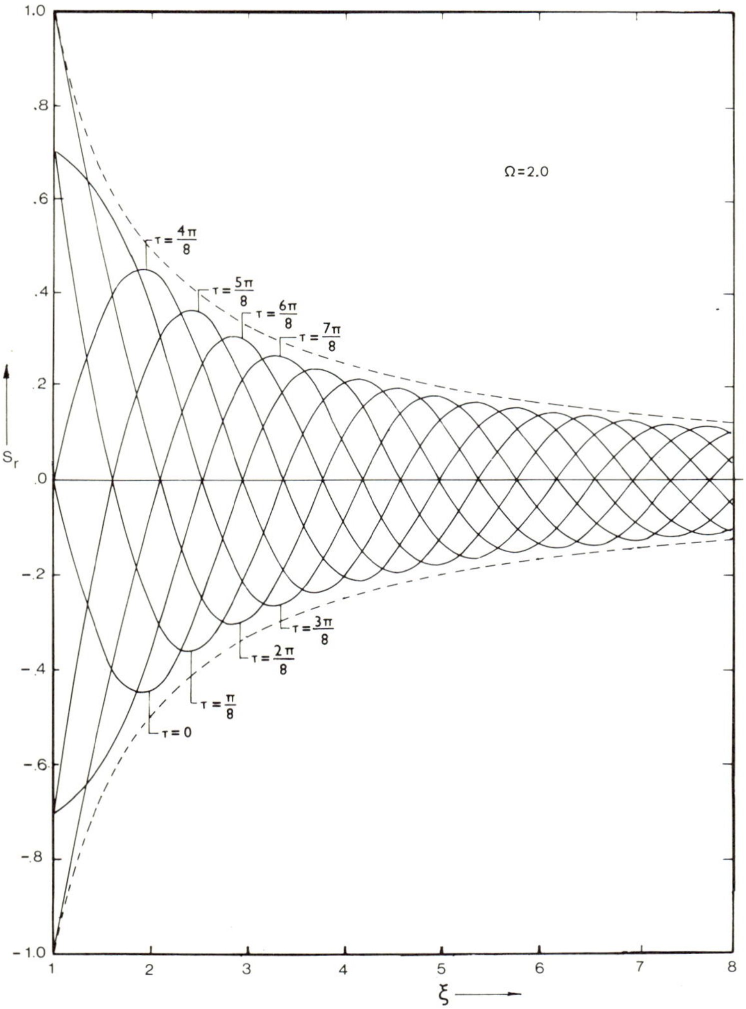

Fig. 37. Stress distribution at different instances within one period in a spherical wave with a harmonic input pulse. Imposed frequency: $\Omega = 2$.

$$S_r(\xi, \tau) = f_1(\xi) \sin(\Omega T) + f_2(\xi) \cos(\Omega T)$$

$$+ \left[f_3(\xi) \sin\left(\beta \frac{T}{A}\right) + f_4(\xi) \cos\left(\beta \frac{T}{A}\right) \right] e^{-T/A} \qquad [5.27]$$

where β and T are defined as before in 3.44 and where 3.45 and 3.46 define the functions f_1, ... f_4 with Ω substituted for ω_1.

It is seen that the solution in 5.27 consists of two parts, one which dies out exponentially with time and one which persists. The first one is called the transient

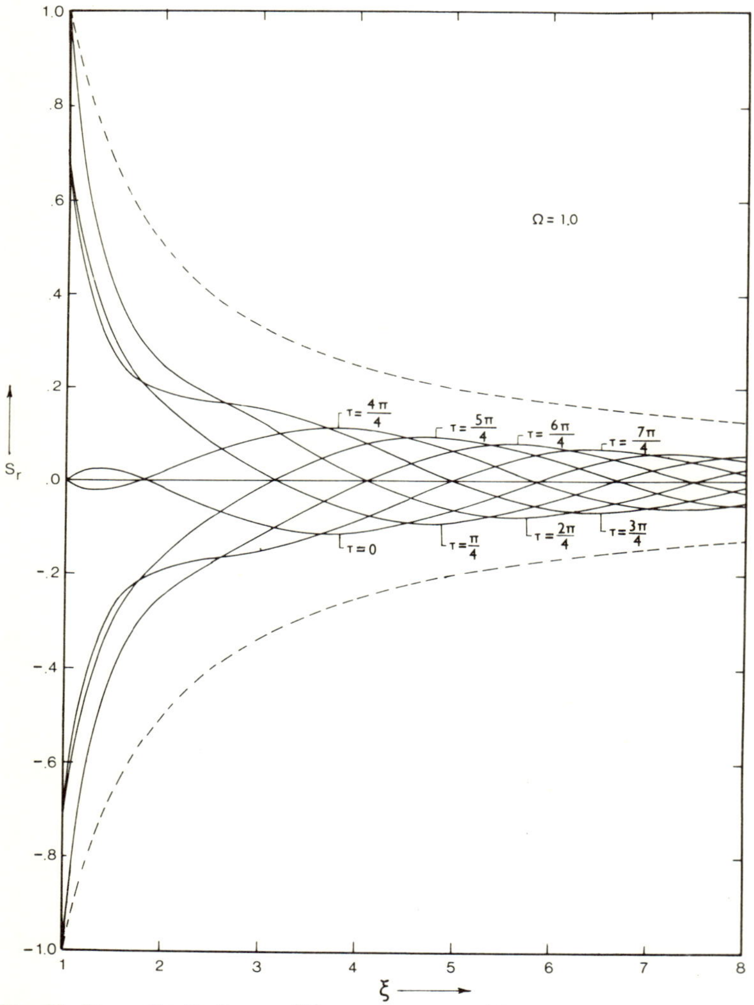

Fig. 38. Stress distribution at different instances within one period in a spherical wave with a harmonic input pulse. Imposed frequency: $\Omega = 1$.

and represents the initial disturbance due to the way in which the excitation is initially applied. The second one is the final semi-steady state, and is exhibited in Figs. 37 and 38.

In Fig. 37 the imposed frequency $\Omega = 2$ and stress S_r is plotted as a function of ξ for $\tau = 0, \pi/8, 2\pi/8, 3\pi/8, 4\pi/8, 5\pi/8, 6\pi/8, 7\pi/8$. The result may be interpreted as a moving train of waves (travelling wave), the maximum amplitudes of which decrease with increasing distance. The waves will move with a velocity v_f which varies with distance, but which for large values of ξ asymptotically approaches velocity c of the compression wave. The attenuation of the maximum amplitude of the travelling wave with increasing values of ξ is exhibited in Fig. 37 by the dotted lines, which represents the attenuation curves for an impulsively started shock wave and is given as ξ^{-1}. It is seen that the attenuation is greater in the present case, but it is easy to show that, for larger values of ξ, the attenuation will also in the present case be proportional to ξ^{-1}.

Fig. 38 shows exactly the same as Fig. 37, the only difference being that now $\Omega = 1.0$ and consequently the curves are plotted for $\tau = 0, \pi/4, 2\pi/4, 3\pi/4, 4\pi/4, 5\pi/4, 6\pi/4$ and $7\pi/4$. It is observed that the initial attenuation of the maximum amplitudes in this case is much greater than in the former. Thus, the imposed frequency has a great influence on the attenuation. However, also in this case the attenuation for larger values of ξ is proportional to ξ^{-1}. The situation therefore is analogous to the one explained in Fig. 23. The velocity of the travelling wave (phase velocity) for large values of ξ is the same as in the first case because it is independent of the imposed frequency.

Chapter 6

CHARACTERISTIC FEATURES OF PLANE WAVES (KELVIN MEDIUM)

INTRODUCTION

The results in connection with the propagation of stress waves in an elastic medium exhibited in the previous chapters have some features, which are of importance when discussing shock waves in rocks. However, they also have features which are intolerable when confronted with results obtained in real materials. As pointed out in Chapter 1, section *The phenomenological relation (Kelvin material),* this situation may be remedied by the introduction of a "damping" effect exhibited by the material, and the Kelvin material in its simplest form presents itself as one of the easiest choices when the behavior of the material is to be generalized. Thus, the present and the following chapters are dedicated to exhibiting the way in which the previous results will be modified by this change in the behavior of the material.

BASIC EQUATIONS

Just as in Chapter 2 the plane wave in a Kelvin medium halfspace will be studied under the assumption that the free surface of the halfspace is given by $x = 0$ in a Cartesian coordinate system (x, y, z). The displacement components, v and w (in the y- and z-direction, respectively) will be assumed equal to zero everywhere, leaving only u different from zero. The concept of a plane wave carries in it the consequence that:

$$\frac{\partial u}{\partial y} = \frac{\partial u}{\partial z} = 0 \qquad [6.1]$$

With this information it is now relatively easy to deduce from 1.37 and 1.38 the following relations, which will exist in this case:

$$\sigma_x = E_1\left(1 + \eta\frac{\partial}{\partial t}\right)\frac{\partial u}{\partial x} \qquad [6.2]$$

and:

$$\tau_{xy} = \tau_{xz} = \tau_{yz} = 0 \qquad [6.3]$$

where:

$$\left.\begin{aligned} E_1 &= \tfrac{1}{3}(k_0 + 2k_d) \\ E_1\eta &= \tfrac{1}{3}(k_0\eta_0 + 2k_d\eta_d) \end{aligned}\right| \qquad [6.4]$$

The equations of motion [1.43] will reduce to:

$$\frac{\partial \sigma_x}{\partial x} = \rho \frac{\partial^2 u}{\partial t^2} \qquad [6.5]$$

which by introduction of eq. 6.2 will convert into:

$$E_1\left(1 + \eta \frac{\partial}{\partial t}\right) \frac{\partial^2 u}{\partial x^2} = \rho \frac{\partial^2 u}{\partial t^2} \qquad [6.6]$$

Before attempting to establish any solution to this equation, non-dimensional quantities should be introduced. It will then be observed, that the problem does not contain any explicit length with which the independent variable x can be made dimensionless. The concept of a plane wave can namely only be maintained if the load applied to the free surface $x = 0$ is evenly distributed over the surface at all times, i.e. the input pulse is a function of time only. As in Chapter 2, it will therefore be assumed that the load at the free surface $x = 0$ may be written as:

$$p = p_0 f(t) \qquad [6.7]$$

where $f(t)$ has a maximum value equal to 1 (leaving p_0 equal to the maximum value of the applied load).

A dimensional analysis of the problem renders the following possibility for non-dimensional quantities:

$$\left.\begin{aligned} &\text{Non-dimensional distance:} && \xi = \frac{x}{\eta}\sqrt{\frac{\rho}{E_1}} \\ &\text{Non-dimensional time:} && \tau = \frac{t}{\eta} \\ &\text{Non-dimensional displacement:} && \Phi = \frac{u}{\eta p_0}\sqrt{E_1\rho} \\ &\text{Non-dimensional stress:} && S = \frac{\sigma_r}{p_0} \end{aligned}\right| \qquad [6.8]$$

If these quantities are introduced into the basic equations the problem may be stated as follows (from eqs. 6.2 and 6.6):

$$S = \left(1 + \frac{\partial}{\partial \tau}\right) \frac{\partial \Phi}{\partial \xi} \qquad [6.9]$$

$$\left(1 + \frac{\partial}{\partial \tau}\right) \frac{\partial^2 \Phi}{\partial \xi^2} = \frac{\partial^2 \Phi}{\partial \tau^2} \qquad [6.10]$$

The initial and boundary conditions for $S(\xi, \tau)$ and $\Phi(\xi, \tau)$ will be:

$$\left.\begin{aligned} &\Phi(\xi, 0) = 0 \\ &\left.\frac{\partial \Phi}{\partial \tau}\right|_{\tau=0} = 0 \end{aligned}\right| \qquad [6.11]$$

$$\left.\begin{aligned} &S(0, \tau) = f(\tau) \\ &S(\infty, \tau) = 0 \end{aligned}\right| \qquad [6.12]$$

It is noticed that application of the operator $\partial/\partial\xi + \partial^2/\partial\xi\partial\tau$ to 6.10 will in view of 6.9 convert the equation into:

$$\frac{\partial^2 S}{\partial \xi^2} + \frac{\partial^2 S}{\partial \tau \partial \xi^2} = \frac{\partial^2 S}{\partial \tau^2} \qquad [6.13]$$

Similarly the initial conditions may be expressed in terms of $S(\xi, \tau)$ as:

$$\left.\begin{aligned} &S(\xi, 0) = 0 \\ &\left.\frac{\partial S}{\partial \tau}\right|_{\tau=0} = 0 \end{aligned}\right| \qquad [6.14]$$

One has in this way arrived at a complete, mathematical formulation of the problem, the only remaining information being the specification of the input pulse $f(\tau)$.

THE HEAVISIDE FUNCTION AS AN INPUT PULSE

If it is assumed that at time $\tau = 0$ a constant normal stress p_0 is suddenly applied to the free surface of the halfspace and then kept constant for all times, the function $f(\tau)$ is specified as a Heaviside function:

$$f(\tau) = \begin{cases} 0 & \tau < 0 \\ 1 & \tau > 0 \end{cases} \qquad [6.15]$$

With this equation all information necessary for a solution is present in eqs. 6.12, 6.13 and 6.14. The solution is sought by means of a Laplace transform applied to 6.13, which, in view of the initial conditions [6.14], will give:

$$\frac{d^2S^*}{d\xi^2}(1+s) = s^2 S^* \qquad [6.16]$$

This equation is to be solved with the boundary condition obtained from 6.12 which, in view of 6.15, will be:

$$\left.\begin{aligned} S^*(0) &= \frac{1}{s} \\ S^*(\infty) &= 0 \end{aligned}\right| \qquad [6.17]$$

The solution is easily found:

$$S^* = \frac{1}{s}\exp\left(-\frac{s\xi}{\sqrt{1+s}}\right) \qquad [6.18]$$

whereupon the inverse Laplace transform provides the formal solution:

$$S(\xi, \tau) = \frac{1}{2\pi i}\int_{b-i\infty}^{b+i\infty} \frac{1}{s}\exp\left(-\frac{s\xi}{\sqrt{1+s}}\right) e^{s\tau}\, ds \qquad [6.19]$$

This integral is not so easily solved by means of deformation of the path of integration. (This is mainly due to the fact that the integrand possesses an essential singularity at $s = -1$.) The integral has, therefore, been evaluated numerically along the indicated path by putting:

$$s = x - 1 + iy \qquad [6.20]$$

and then letting y vary from $-\infty$ to $+\infty$ for a fixed value of x greater than 1. Under these circumstances the argument of the exponential function in the integrand will be:

$$s\tau - s\xi(1+s)^{-\frac{1}{2}} = H + iK \qquad [6.21]$$

where:

$$H = (x - 1)\,\tau - \xi\left(1 - \frac{1}{\rho}\right)\sqrt{\frac{\rho + x}{2}}$$
$$K = y\tau - \xi\left(1 + \frac{1}{\rho}\right)\sqrt{\frac{\rho - x}{2}} \qquad [6.22]$$

with:

$$\rho = \sqrt{x^2 + y^2} \qquad [6.23]$$

One will then find that eq. 6.19 may be converted into the following ordinary integral:

$$S(\xi, \tau) = \frac{1}{\pi}\int_0^\infty \frac{e^{+H(y)}}{(x-1)^2 + y^2}\,[(x-1)\cos K + y \sin K]\,dy \qquad [6.24]$$

which then can easily be evaluated numerically. The result of this evaluation is shown in the diagrams of Figs. 39 and 40.

In Fig. 39 the solution is exhibited for different values of time τ as a function of ξ, i.e. the diagram illustrates the state of stress in the entire field at particular times. It is seen that the characteristic feature of the plane wave in the elastic medium with a sharp shock front is completely changed. Instead, a gradual increase of the stress occurs at each point. To illustrate this, attention is drawn to Fig. 40 where the solution is exhibited as a function of time for discrete positions ξ. The sudden rise in stress, usually connected with a shock wave, occurs only at extremely small values of ξ. However, for larger values of ξ it is observed that a considerable time elapses before anything seems to happen. Then, when measurable stresses are created, they develop slowly as seen for $\xi = 5$ in Fig. 40.

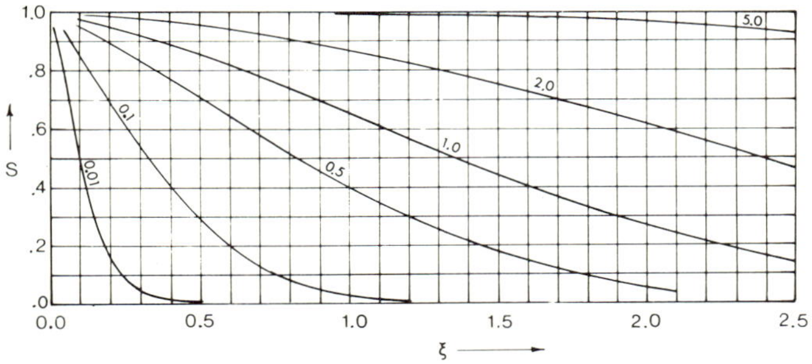

Fig. 39. Distribution of the stress, S, as a function of ξ at different instances, τ (plane wave, Heaviside input pulse).

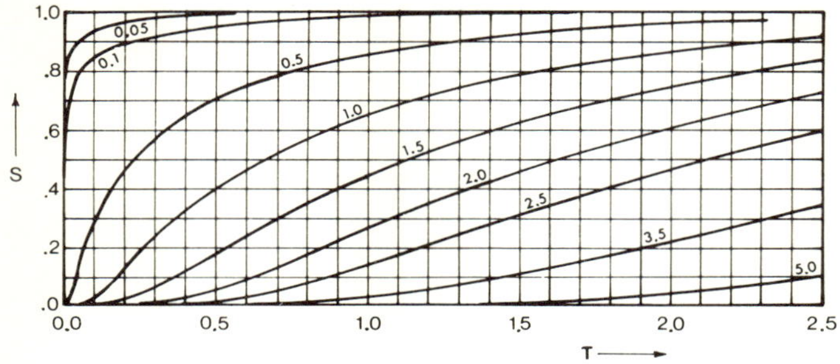

Fig. 40. Stress S at different positions, ξ, exhibited as a function of time, τ (plane wave, Heaviside input pulse).

The sharp discontinuity in the input pulse gave rise to a moving discontinuity (the shock front) in the elastic medium. In the present case this feature disappears due to the introduced damping. The process of propagation of the shock wave becomes here more similar to the propagation of a viscous wave in a Newtonian fluid. This is, of course, caused by the fact that the "viscosity" is introduced in analogous ways in the two cases.

Finally it should be observed that no explicit velocity of propagation corresponding to the velocity c of a compression wave or velocity c_t of a shear wave can be deduced directly from the basic equations. Looking at Fig. 39 one might be tempted to say that the low stress, e.g. $S = 0.05$, is propagated at a higher speed than a higher stress, e.g. $S = 0.95$, which, however, is just another way of expressing the asymptotic behavior of the solution.

THE RECTANGULAR INPUT PULSE

If the input pulse acting on the free surface ($x = 0$) of the halfspace is rectangular, i.e. if the suddenly applied stress p_0 is suddenly removed after a non-dimensional time τ_0, one has a rectangular input pulse. In this case $f(\tau)$ is to be specified as:

$$f(\tau) = \begin{cases} 0 & \tau < 0 \\ 1 & 0 < \tau < \tau_0 \\ 0 & \tau > \tau_0 \end{cases} \qquad [6.25]$$

It may be conceived of as a superposition of two Heaviside input pulses which are displaced τ_0 in time and are of opposite signs. Thus the solution given in the previous section may be used to directly establish the solution for the present case. This solution is presented in the form of tables in Persen (1968), and Figs. 41,

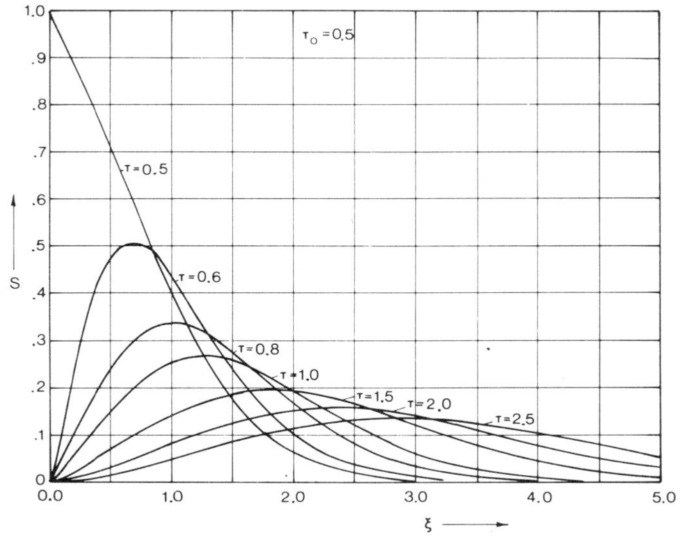

Fig. 41. Stress S at different instances, τ, plotted as a function of ξ for τ_0 = 0.5 (plane wave, rectangular input pulse).

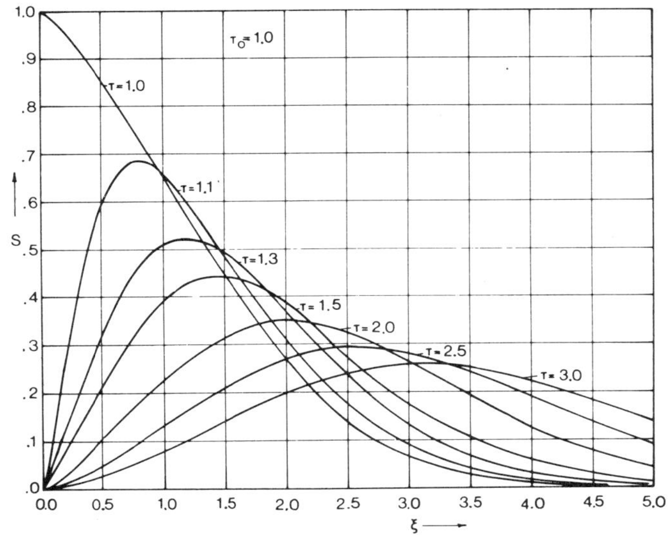

Fig. 42. Stress S at different instances, τ, plotted as a function of ξ for τ_0 = 1.0 (plane wave, rectangular input pulse).

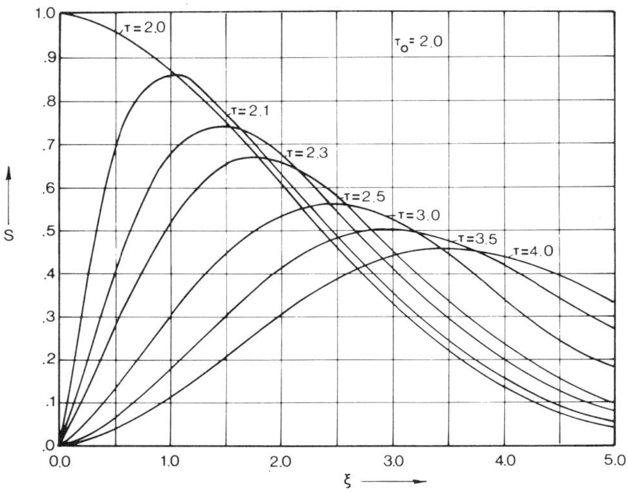

Fig. 43. Stress S at different instances, τ, plotted as a function of ξ for τ_0 = 2.0 (plane wave, rectangular input pulse).

42 and 43 show the distribution of stress in the medium at different instances τ for pulse durations τ_0 = 0.5, 1.0 and 2.0, respectively.

(1) It is first observed that in contrast to the same case in an elastic medium, an attenuation of the maximum amplitude takes place as the wave propagates.

(2) Secondly it is observed that the attenuation is strongly dependent on the duration τ_0 of the input pulse.

Because this behavior is much more realistic than in the elastic case, it may be worthwhile to exhibit this in a diagram. Fig. 44 shows the maximum amplitude, S_{max}, of the wave as a function of ξ for the three different values of τ_0 in the previous figures. It shows clearly that for larger values of ξ the curves in the doubly logarithmic diagram become straight lines with the same slope. The slopes of the lines may be used as a measure of the attenuation of S_{max}, but it is seen how strongly the vertical position of these straight lines is influenced by τ_0. Because of the rather unrealistic geometry of the present problem, further comments are not being made at this point, but it should be emphasized that exactly such considerations as those presented here on the attenuation of S_{max} will play an important role in the application of theoretical results to experimental data obtained in rocks.

THE FINITE MEDIUM

The problem discussed here is identical to the one examined in Chapter 2, section *Plane waves in the halfspace*, with the only exception that here the material is

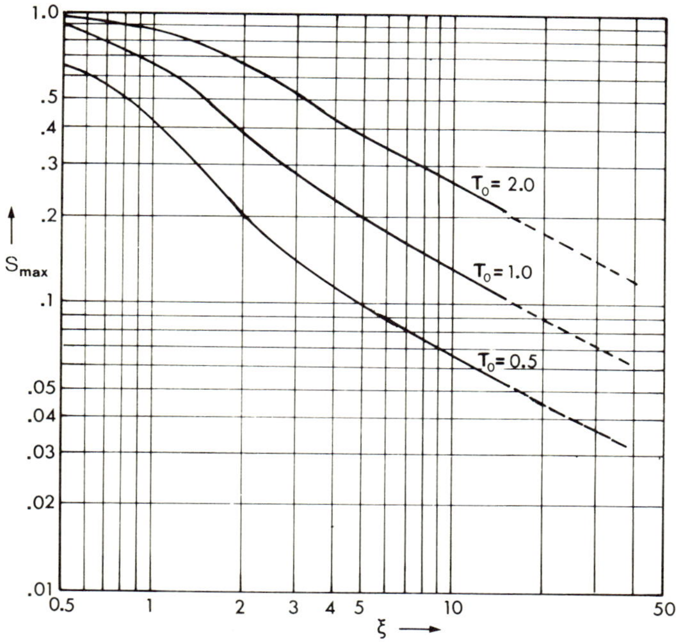

Fig. 44. S_{max} as a function of distance ξ. (Attenuation curve, rectangular input pulse.)

assumed to be of the Kelvin type in the sense previously described. The slab has a finite thickness $2h$ (Fig. 7), and thus the problem will now contain the finite length, which was missing in the previous problem, and which now may be used to non-dimensionalize the physical quantities. Because the waves to be considered are still plane, the basic equations [6.2 and 6.6] are still valid, but now the non-dimensional quantities will be

Dimensionless distance: $\xi = \dfrac{x}{h}$

Dimensionless time: $\tau = \dfrac{t}{h}\sqrt{\dfrac{E_1}{\rho}}$

Dimensionless displacement: $\Phi = \dfrac{uE_1}{p_0 h}$ [6.26]

Dimensionless stress: $S = \dfrac{\sigma_x}{p_0}$

Dimensionless damping parameter: $\kappa = \dfrac{\eta}{h}\sqrt{\dfrac{E_1}{\rho}}$

It is seen that now for the first time a new parameter, κ, enters the problem, a parameter which will play an important role in the intended practical application. With these non-dimensional quantities introduced, the basic equations (eqs. 6.2 and 6.6) will be transformed into:

$$S = \left(1 + \kappa \frac{\partial}{\partial \tau}\right) \frac{\partial \Phi}{\partial \xi} \qquad [6.27]$$

$$\left(1 + \kappa \frac{\partial}{\partial \tau}\right) \frac{\partial^2 \Phi}{\partial \xi^2} = \frac{\partial^2 \Phi}{\partial \tau^2} \qquad [6.28]$$

A few remarks may now be added on the importance of the finite length h in the problem. It is clear that the solution to the differential equation [6.28] will be Φ, given as a function of ξ, τ and parameter κ. Thus:

$$\Phi = \Phi(\xi, \tau, \kappa) = \Phi\left(\frac{x}{h}, \frac{t}{h} \sqrt{\frac{E_1}{\rho}}, \frac{\eta}{h} \sqrt{\frac{E_1}{\rho}}\right) \qquad [6.29]$$

In the absence of a length (h) in a problem, the variables above will have to appear in such a combination, that this length h vanishes from the solution; i.e.:

$$\Phi = \Phi\left(\frac{\xi}{\kappa}, \frac{\tau}{\kappa}\right) = \Phi\left(\frac{x}{\eta} \sqrt{\frac{\rho}{E_1}}, \frac{t}{\eta}\right) \qquad [6.30]$$

It is seen that in this way exactly those non-dimensional quantities appear, which were introduced in the previous section.

Before proceeding to the solution of the problem it is observed that the boundary conditions will be expressed in terms of the non-dimensional stress, S. Therefore the operator $(1 + \kappa\partial/\partial\tau)(\partial/\partial\xi)$ is applied to eq. 6.28, which then converts into:

$$\left(1 + \kappa \frac{\partial}{\partial \tau}\right) \frac{\partial^2 S}{\partial \xi^2} = \frac{\partial^2 S}{\partial \tau^2} \qquad [6.31]$$

This is the differential equation for $S(\xi, \tau)$ which will have to be solved for the following boundary and initial conditions:

$$\left.\begin{aligned} S(1, \tau) &= 1 \\ S(-1, \tau) &= -1 \end{aligned}\right| \qquad [6.32]$$

$$\left.\begin{aligned} S(\xi, 0) &= 0 \\ \left.\frac{\partial S}{\partial \tau}\right|_{\tau=0} &= 0 \end{aligned}\right| \qquad [6.33]$$

One might, as before, proceed to seek the solution by means of a Laplace transform, but because the solution eventually will appear as a Fourier series, a product substitution may as well be introduced:

$$S(\xi, \tau) = 1 + S_1(\xi) \cdot S_2(\tau) \qquad [6.34]$$

In terms of S_1 the boundary conditions will then be:

$$\left.\begin{aligned} S_1(1) &= 0 \\ S_1(-1) &= 0 \end{aligned}\right| \qquad [6.35]$$

Substitution of 6.34 into the differential equation (eq. 6.31) will lead to:

$$\frac{d^2 S_1}{d\xi^2}\left(S_2 + \kappa \frac{dS_2}{d\tau}\right) = S_1 \frac{d^2 S_2}{d\tau^2} \qquad [6.36]$$

This equation can be rearranged in such a way that the right-hand side is a function of ξ only and the left-hand side is a function of τ only. This equation can then only be satisfied if both sides are equal to a constant, e.g. λ^2, whereupon the following two differential equations for S_1 and S_2 are obtained:

$$\frac{d^2 S_1}{d\xi^2} + \lambda^2 S_1 = 0 \qquad [6.37]$$

$$\frac{d^2 S_2}{d\tau^2} + \lambda^2 \kappa \frac{dS_2}{d\tau} + \lambda^2 S_2 = 0 \qquad [6.38]$$

Observing that λ so far is an unknown constant, the solutions to these equations are easily found:

$$\left.\begin{aligned} S_1(\xi) &= A \sin(\lambda\xi) + B \cos(\lambda\xi) \\ S_2(\tau) &= C\, e^{\alpha_1 \tau} + D\, e^{\alpha_2 \tau} \end{aligned}\right| \qquad [6.39]$$

where:

$$\alpha_{1,2} = -\frac{\lambda^2 \kappa}{2} \pm \sqrt{\frac{\lambda^4 \kappa^2 - 4\lambda^2}{4}} \qquad [6.40]$$

With the solution for S_1 given in 6.39 the boundary conditions [6.35] will now appear in the following form:

$$\begin{aligned} A \sin(\lambda) + B \cos(\lambda) &= 0 \\ -A \sin(\lambda) + B \cos(\lambda) &= 0 \end{aligned} \qquad [6.41]$$

This is a set of homogeneous equations for the two constants of integration A and B, and it has only a non-trivial solution in the case when the determinant of the system is equal to zero. In this way λ is determined and it is seen that there exists an infinite number of values for λ for which this is true:

$$\lambda = \lambda_n = \frac{n\pi}{2} \qquad (n = \text{odd integer}) \qquad [6.42]$$

The solution then appears as a sum of all possible values of λ, i.e.:

$$S = 1 + \sum_{n=1}^{\infty} (C_n \, e^{\alpha_1 \tau} + D_n \, e^{\alpha_2 \tau}) \cos\left(\frac{n\pi\xi}{2}\right) \qquad [6.43]$$

where now:

$$\begin{aligned} \alpha_1 &= -\frac{n^2\pi^2\kappa}{8} + \frac{1}{8}\sqrt{n^2\pi^2(n^2\pi^2\kappa^2 - 16)} \\ \alpha_2 &= -\frac{n^2\pi^2\kappa}{8} - \frac{1}{8}\sqrt{n^2\pi^2(n^2\pi^2\kappa^2 - 16)} \end{aligned} \qquad [6.44]$$

Now the initial conditions [6.33] are considered, and the first one of these will give:

$$S(\xi, 0) = 0 = 1 + \sum_{n=1}^{\infty} (C_n + D_n) \cos\left(\frac{n\pi\xi}{2}\right)$$

or: [6.45]

$$\sum_{n=1}^{\infty} (C_n + D_n) \cos\left(\frac{n\pi\xi}{2}\right) = -1$$

Differentiation of 6.43 will give:

$$\frac{\partial S}{\partial \tau} = \sum_{n=1}^{\infty} (C_n \alpha_1 \, e^{\alpha_1 \tau} + D_n \, \alpha_2 \, e^{\alpha_2 \tau}) \cos\left(\frac{n\pi\xi}{2}\right) \qquad [6.46]$$

and the second initial condition [6.33] will be satisfied if:

$$C_n \alpha_1 + D_n \alpha_2 = 0 \qquad [6.47]$$

Eq. 6.45 requires some extra consideration. The sum $(C_n + D_n)$ must be conceived of as the Fourier coefficients of the function -1 between $\xi = -1$ and $\xi = +1$. This means that:

$$C_n + D_n = \frac{4}{n\pi} \sin\left(\frac{n\pi}{2}\right) \qquad [6.48]$$

The two constants C_n and D_n can now easily be solved from eqs. 6.47 and 6.48, whereupon the final solution will appear in the following form:

$$S = 1 - \sum_{n=1}^{\infty} \frac{4}{n\pi} (\alpha_1\, e^{\alpha_2 \tau} - \alpha_2\, e^{\alpha_1 \tau}) \frac{\sin(n\pi/2)}{\alpha_1 - \alpha_2} \cos\left(\frac{n\pi\xi}{2}\right) \qquad [6.49]$$

A few remarks to this solution seem appropriate. It is noticed that the expressions for α_1 and α_2 will have only real values if:

$$\text{(super-critical)}\ \kappa > \frac{4}{n\pi} \qquad [6.50]$$

This means that every term in the series is super-critically damped. The sub-critical

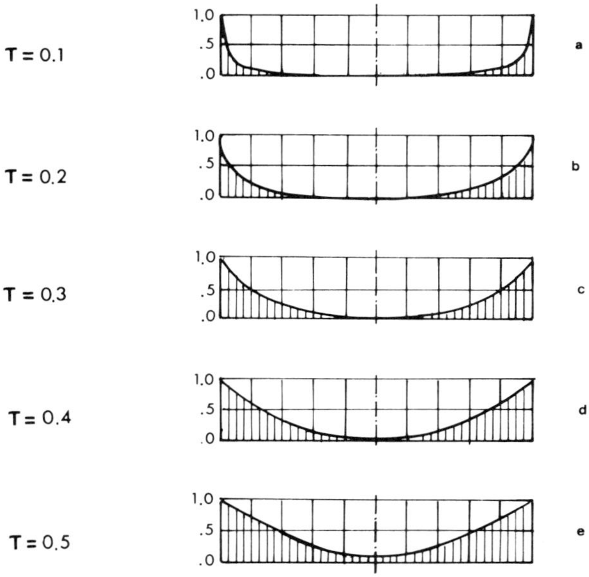

Fig. 45. The development with time τ of the state of stress (S) in the slab for the unit load $f(\tau) = 1$ applied impulsively to the surfaces ($\tau < 0.5$).

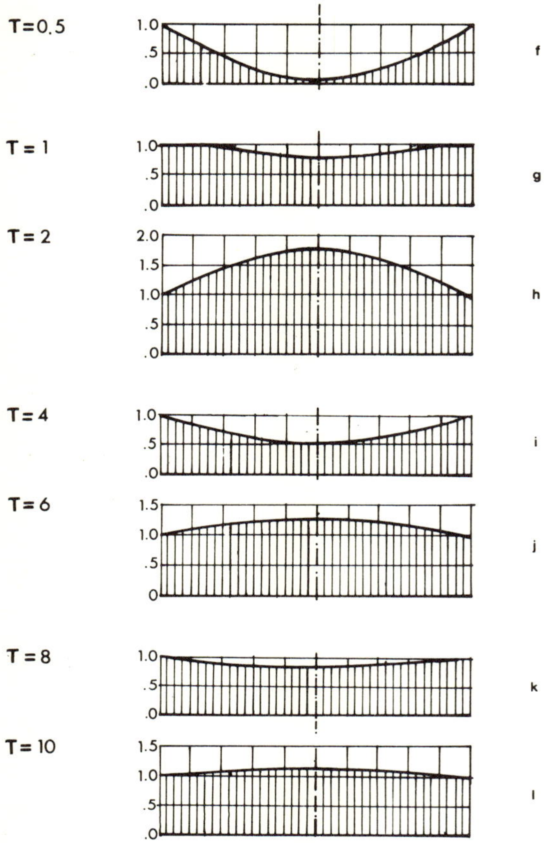

Fig. 46. The development with time τ of the state of stress (S) in the slab for the unit load $f(\tau) = 1$ applied impulsively to the surfaces ($0.5 < \tau < 10$).

damping will appear when:

$$\text{(sub-critical) } \kappa < \frac{4}{n\pi} \qquad [6.51]$$

Thus the following situation will exist:

(1) For $\kappa > 4/\pi$ all terms in the solution eq. 6.49 are super-critically damped.

(2) For $\kappa < 4/\pi$ will at least the first term in the solution be sub-critically damped. The smaller κ is, the larger will the number of sub-critically damped terms in the solution be, but in all cases one will after a finite number of terms reach a point in the series after which all terms are super-critically damped.

One final remark should be made to the solution. When κ is equal to $4/n\pi$, then

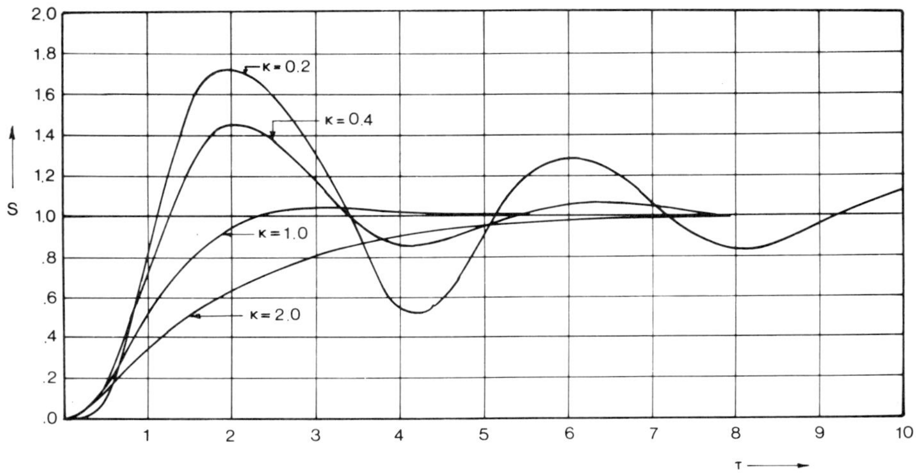

Fig. 47. The stress at the center of the slab as a function of time τ for different values of κ.

$\alpha_1 = \alpha_2$. This means that the term for which this happens will have to be exchanged for the following expression:

$$\left(\frac{4}{n\pi} - 2\tau\right) e^{-n\pi\tau/2} \sin\left(\frac{n\pi}{2}\right) \cos\left(\frac{n\pi\xi}{2}\right) \qquad [6.52]$$

The solution (eq. 6.49) has now been evaluated numerically for $\kappa = 0.4$, and the result is exhibited in Figs. 45 and 46. The initial behavior is shown in Fig. 45 where the absence of a sharp shock front (as compared with the elastic case, see Fig. 8) is again obvious. Fig. 46 shows the further development and it is noticed that the state of stress where a uniform constant stress $S = 1$ persists everywhere will be reached rather rapidly. The influence of the damping parameter, κ, on this development may be better exhibited by observing how the stress in the middle of the slab develops with time for different values of κ. This is done in Fig. 47. For lower values of κ this stress is seen to oscillate several times before reaching its final value 1. For larger values of κ the stress increases monotonically towards the same value. It is realized that the present solution seems to give a much more realistic picture of the physical realities than the elastic case.

Chapter 7

THE SPHERICAL WAVE (KELVIN MEDIUM)

INTRODUCTION

The present chapter will deal with the same problems as those originally described in Chapter 3, the main difference being the damping effect exhibited by the Kelvin medium as compared with the elastic medium. It was stressed before, that the solution to the spherically symmetric wave has an immediate application to the propagation of stress waves created in rocks by confined detonating charges. The main emphasis in the exposition will thus be to exhibit those features of the solution which have a direct bearing on the interpretation of experimental results.

BASIC EQUATIONS

For a spherical wave the kinematic conditions expressed in Chapter 3, section *Basic equations,* will still be valid, because they do not depend on the process or the material. Thus the equation of motion may be taken from 3.5 and the expressions for the strain from 3.3. The expressions for the relation between the state of stress and the state of strain is in the present case obtained from 1.35 which gives:

$$I_1 = \sigma_r + \sigma_\varphi + \sigma_\theta = k_0 \left(1 + \eta_0 \frac{\partial}{\partial t}\right) \left(\frac{\partial v_r}{\partial r} + \frac{2v_r}{r}\right) \qquad [7.1]$$

and:

$$\left.\begin{aligned} \sigma_r - \tfrac{1}{3} I_1 &= k_d \left(1 + \eta_d \frac{\partial}{\partial t}\right) \left[\frac{\partial v_r}{\partial r} - \frac{1}{3}\left(\frac{\partial v_r}{\partial r} + \frac{2v_r}{r}\right)\right] \\ \sigma_\varphi - \tfrac{1}{3} I_1 &= k_d \left(1 + \eta_d \frac{\partial}{\partial t}\right) \left[\frac{v_r}{r} - \frac{1}{3}\left(\frac{\partial v_r}{\partial r} + \frac{2v_r}{r}\right)\right] \\ \sigma_\theta - \tfrac{1}{3} I_1 &= k_d \left(1 + \eta_d \frac{\partial}{\partial t}\right) \left[\frac{v_r}{r} - \frac{1}{3}\left(\frac{\partial v_r}{\partial r} + \frac{2v_r}{r}\right)\right] \end{aligned}\right| \qquad [7.2]$$

It is a matter of routine algebra to express stresses σ_r, σ_φ and σ_θ explicitly from these equations, whereupon introduction into the equation of motion [3.5] will give:

$$E\left(1+\eta\frac{\partial}{\partial t}\right)\left[\frac{\partial^2 v_r}{\partial r^2}+\frac{2}{r}\frac{\partial v_r}{\partial r}-\frac{2v_r}{r^2}\right]=\rho\frac{\partial^2 v_r}{\partial t^2} \tag{7.3}$$

where E_1 and η are determined as before in 6.4.

Referring again to Chapter 3, the spherical wave is assumed to be created by an input pulse p given in 3.1 and acting on the inside wall of a spherical cavity of a radius a. Thus the problem contains two parameters, p_0 and a, in addition to those appearing in the differential equation [7.3]. The following non-dimensional quantities may thus be introduced.

Non-dimensional distance: $\xi = \dfrac{r}{a}$

Non-dimensional time: $\tau = \dfrac{t}{a}\sqrt{\dfrac{E_1}{\rho}}$

Non-dimensional displacement: $\Phi = \dfrac{E_1 v_r}{p_0 a}$ [7.4]

Non-dimensional stress: $S_r = \dfrac{\sigma_r}{p_0}$

Non-dimensional damping parameter: $\kappa = \dfrac{\eta}{a}\sqrt{\dfrac{E_1}{\rho}}$

With these non-dimensional quantities introduced into 7.3 the differential equation becomes:

$$\left(1+\kappa\frac{\partial}{\partial\tau}\right)\left(\frac{\partial^2\Phi}{\partial\xi^2}+\frac{2}{\xi}\frac{\partial\Phi}{\partial\xi}-\frac{2}{\xi^2}\Phi\right)=\frac{\partial^2\Phi}{\partial\tau^2} \tag{7.5}$$

The boundary and initial conditions for this equation are, however, expressed in terms of stress $S_r(\xi,\tau)$:

$$S_r(1,\tau)=f(\tau)$$

$$S_r(\infty,\tau)=0$$

$$S_r(\xi,0)=0 \tag{7.6}$$

$$\left.\frac{\partial S_r}{\partial\tau}\right|_{\tau=0}=0$$

Thus the relation between stress σ_r and the components of strain must be found from 7.1 and 7.2:

$$\sigma_r = k_d \left(1 + \eta_d \frac{\partial}{\partial t}\right) \frac{\partial v_r}{\partial r} + \tfrac{1}{3}\left[(k_0 - k_d) + (k_0\eta_0 - k_d\eta_d)\frac{\partial}{\partial t}\right]\left(\frac{\partial v_r}{\partial r} + \frac{2v_r}{r}\right) \quad [7.7]$$

In this expression all four constants (k_0, k_d, η_0 and η_d) appear, and these cannot be replaced by the constants E_1 and η without further comments. It is easily shown that introduction of 7.4 will convert 7.7 into:

$$S_r = \left(1 + \kappa \frac{\partial}{\partial \tau}\right)\frac{\partial \Phi}{\partial \xi} + \left(1 - \frac{k_d}{E_1}\right)\left(1 + \frac{E_1\eta - k_d\eta_d}{(E_1 - k_d)\eta}\kappa \frac{\partial}{\partial \tau}\right)\frac{2\Phi}{\xi} \quad [7.8]$$

Now, one may introduce ν_1 as a Kelvin-material analogy to the Poisson's ratio in the elastic material, defined by:

$$1 - \frac{k_d}{E_1} = \frac{\nu_1}{1 - \nu_1} \quad [7.9]$$

In addition one may further simplify the problem by assuming:

$$\eta_0 = \eta_d = \eta \quad [7.10]$$

This implies the assumption, that the dissipation during a change of volume at constant shape is equal to that occurring during a change of shape at constant volume. Such an assumption may in many cases be objectionable, but at this stage the scope of the analysis may be drawn into consideration. For stress-wave propagation in rocks the above assumption is a fair one because of the physical properties of the rock as it occurs in nature with its cracks and other discontinuous surfaces. It should, however, be realized that, for future sophistication (or for other purposes), this assumption may be reconsidered without introducing insurmountable difficulties.

With these remarks, the final form of 7.8 will now be:

$$S_r = \left(1 + \kappa \frac{\partial}{\partial \tau}\right)\left(\frac{\partial \Phi}{\partial \xi} + \frac{\nu_1}{1 - \nu_1}\frac{2\Phi}{\xi}\right) \quad [7.11]$$

and the mathematical statement of the problem consists of the three equations 7.5, 7.6 and 7.11.

The solution is as usual sought by means of a Laplace transform. Applied to 7.5 the transformation gives:

$$\frac{\partial^2 \Phi^*}{\partial \xi^2} + \frac{2}{\xi}\frac{\partial \Phi^*}{\partial \xi} - \left(\frac{2}{\xi^2} + \frac{s^2}{1 + \kappa s}\right)\Phi^* = 0 \quad [7.12]$$

which is the same equation as 3.12 with s^2 replaced by $s^2/(1 + \kappa s)$. This means that the previous solution may be used with:

$$\zeta = \frac{s}{\sqrt{1 + s\kappa}} \qquad [7.13]$$

whereupon:

$$S_r^*(\xi, s) = \frac{[A_1(\zeta\xi)^2 + 2(\zeta\xi) + 2]\, f^*(s)}{[A_1\zeta^2 + 2\zeta + 2]\, \xi^3}\, e^{-\zeta(\xi-1)} \qquad [7.14]$$

Here:

$$A_1 = \frac{1 - \nu_1}{1 - 2\nu_1} \qquad [7.15]$$

i.e. A_1 is defined by ν_1 in the same way as A was defined by ν in 3.19.

THE HEAVISIDE INPUT PULSE

The input pulse defined as a Heaviside function will give:

$$f^*(s) = \frac{1}{s} \qquad [7.16]$$

Then all quantities in 7.14 are known and the inverse Laplace transform will give the expression for $S_r(\xi, \tau)$. The inversion problem is treated in Appendix I and the solution is therefore quoted here with reference to this procedure.

The function $F(A_1, \zeta)$ is here defined as:

$$F(A_1, \zeta) = \frac{[A_1(\zeta\xi)^2 + 2(\zeta\xi) + 2]}{(A_1\zeta^2 + 2\zeta + 2)\, \xi^3}\, e^{-\zeta(\xi-1)} \qquad [7.17]$$

The real and imaginary part of this function for $\zeta = ia$:

$$\left.\begin{aligned} \Psi_r(a) &= K \cos[a(\xi - 1)] - H \sin[a(\xi - 1)] \\ \Psi_i(a) &= K \sin[a(\xi - 1)] + H \cos[a(\xi - 1)] \end{aligned}\right| \qquad [7.18]$$

where:

$$\left.\begin{aligned} K(a) &= \frac{(2 - A_1a^2)(2 - A_1a^2\xi^2) + 4a^2\xi}{[(2 - A_1a^2)^2 + 4a^2]\, \xi^3} \\ H(a) &= -\frac{2a(\xi - 1)(2 + A_1a^2\xi)}{[(2 - A_1a^2)^2 + 4a^2]\, \xi^3} \end{aligned}\right| \qquad [7.19]$$

The final stationary stage is here given by:

$$S_r(\xi, \infty) = \frac{1}{\xi^3} \qquad [7.20]$$

The final solution will then according to eq. A.20 (see Appendix I) be:

$$S_r(\xi, \tau) = \frac{1}{2\xi^3} - \frac{1}{\pi}\int_{1/\kappa}^{\infty} \frac{e^{-(\rho+1/\kappa)\tau}}{\rho + \frac{1}{\kappa}} \{K(a) \sin [a(\xi - 1)] + H(a) \cos [a(\xi - 1)]\} d\rho$$

$$+ \frac{1}{\pi}\int_0^{\pi} \frac{e^{-\tau(1-\cos\varphi)/\kappa} \sin\left(\frac{\varphi}{2}\right)}{1 - \cos\varphi} \left\{K(b) \sin\left[\frac{\tau}{\kappa} \sin\varphi - b(\xi - 1) + \frac{\varphi}{2}\right] - \right.$$

$$\left. - H(b) \cos\left[\frac{\tau}{\kappa} \sin\varphi - b(\xi - 1) + \frac{\varphi}{2}\right]\right\} d\varphi \qquad [7.21]$$

where $a(\rho)$ and $b(\varphi)$ are defined in eqs. A.7 and A.10, respectively.

The complexity of this solution makes it unrewarding to try to bring out its different features analytically. The solution has therefore been evaluated numerically for different values of κ and ξ as a function of τ and the results are to be found in Persen (1968). These tables will now be used as a basis for the further discussion.

THE ARBITRARY INPUT PULSE

The solution established in the previous section is obtained on the condition that the input pulse $f(\tau)$ is a Heaviside function. It is then easily found that the solution to a rectangular input pulse can be obtained from the above solution as the difference between two such solutions which are displaced in time by an amount τ_0 equal to the pulse duration of the rectangular pulse. This was exhibited in the cases of an elastic medium (Chapters 3, 4 and 5). However, if the input pulse is of arbitrary shape, the input pulse may be replaced by a superposition of a number of rectangular pulses as sketched in Fig. 48. This will in the limit lead to a Duhamel integral representation of the solution. In very many cases, however, the exact form of the input pulse is not known. This is especially true when stress waves in rocks created by explosions are to be described. It is thus essential to bring out some general features of the solution which makes the theory more readily applicable.

First the attention is drawn to the exposition of the spherical wave in an elastic medium in Chapter 3. It was shown how the maximum amplitude of stress S_r created at given locations ξ by the input pulse would be attenuated with increasing values of ξ. The influence on this attenuation of the "rising time" of the input pulse

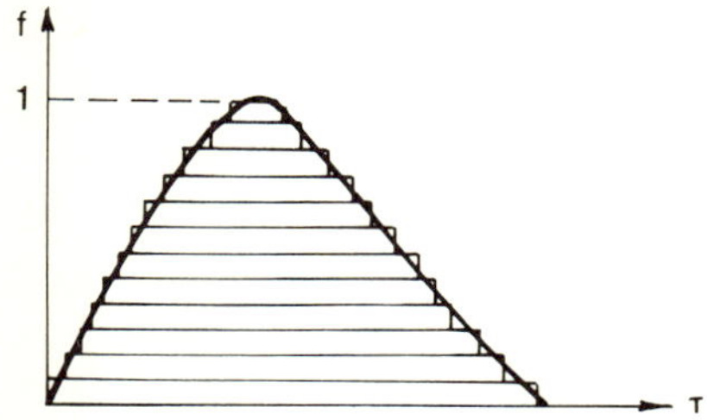

Fig. 48. The arbitrary input pulse and its approximation.

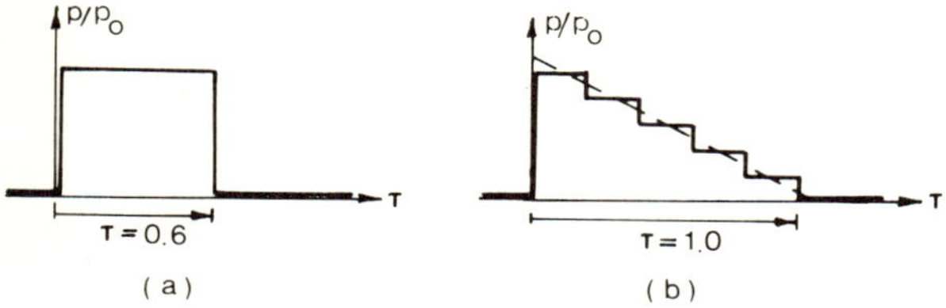

Fig. 49. The input pulse: (a) rectangular, (b) approximate triangular.

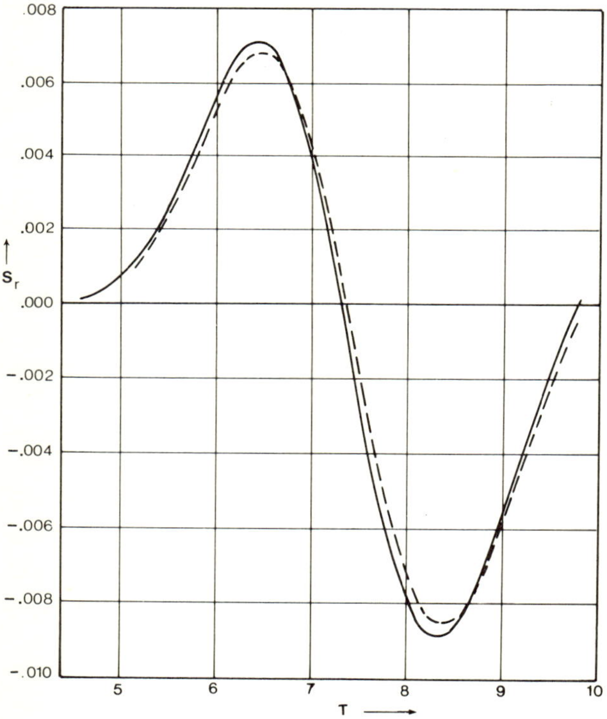

Fig. 50. Stress S_r plotted as a function of time τ at $\xi = 8$ for a rectangular pulse (solid curve) and a triangular pulse (dotted curve) ($\kappa = 0.125$).

TABLE II

Comparison of maximum S_r-values

κ	ξ	Rectangular $S_{r,\max}$	Triangular $S_{r,\max}$	Ratio rectangular/ triangular
0.020	2	0.36153	0.30723	1.177
	9	0.02605	0.02302	1.131
	17	0.00748 8	0.00689 5	1.086
0.125	8	0.00715 4	0.00684 2	1.046
	17	0.00113 9	0.00111 4	1.023
0.175	7	0.00725 9	0.00697 3	1.041
	20	0.00051 10	0.00050 34	1.051

and/or its impulse was touched upon. A similar investigation may now be carried out for the two input pulses sketched in Fig. 49. They are both characterized by an impulsive start at $\tau = 0$ at which time they obtain their maximum value 1. The first pulse (Fig. 49a) is a rectangular one with a duration time $\tau_0 = 0.6$, the second one (Fig. 49b) is a step function approximation to a triangular pulse. The impulse of the input pulses is $I = 0.6$ in both cases. Stress S_r created at $\xi = 8$ is plotted in Fig. 50 as a function of time τ with $\kappa = 0.125$. This will be denoted "the signal".

First it is observed that no trace of the discontinuities of the input pulses can be observed in the signal in Fig. 50. This is due to the influence of the damping effect in the material. Secondly it is seen that the signals created by the two input pulses are almost identical. Attention is especially drawn to the maximum value of the first positive phase of the signal which deviates less than 5% from each other. This would be significant if it could be shown to be a general feature. Table II gives the computed maximum values of S_r at different locations (ξ) and for different values of the damping parameter κ. The values are obtained from the assumed input pulses in Fig. 49. It is seen that the maximum values of S_r created by the two pulses differ by a percentage amount which decreases with increasing distance (ξ) and increasing values of the damping parameter κ. In view of the range of values for ξ and κ, which will be of importance when shock waves in rocks are concerned, the following statement may be made:

Within an accuracy of 15% (and within the range of $5 \leqslant \xi \leqslant 20$) the maximum values of S_r created by an arbitrary input pulse, which starts impulsively, will be the same as the ones created by a rectangular pulse whose duration time, τ_0, is determined such that its impulse is equal to that of the arbitrary pulse.

THE ATTENUATION CURVES

The preceding section has placed the emphasis on the maximum value of stress S_r and how this value decreases with increasing values of distance ξ. The damping effect is expressed through the parameter κ, and the stress wave will in this case

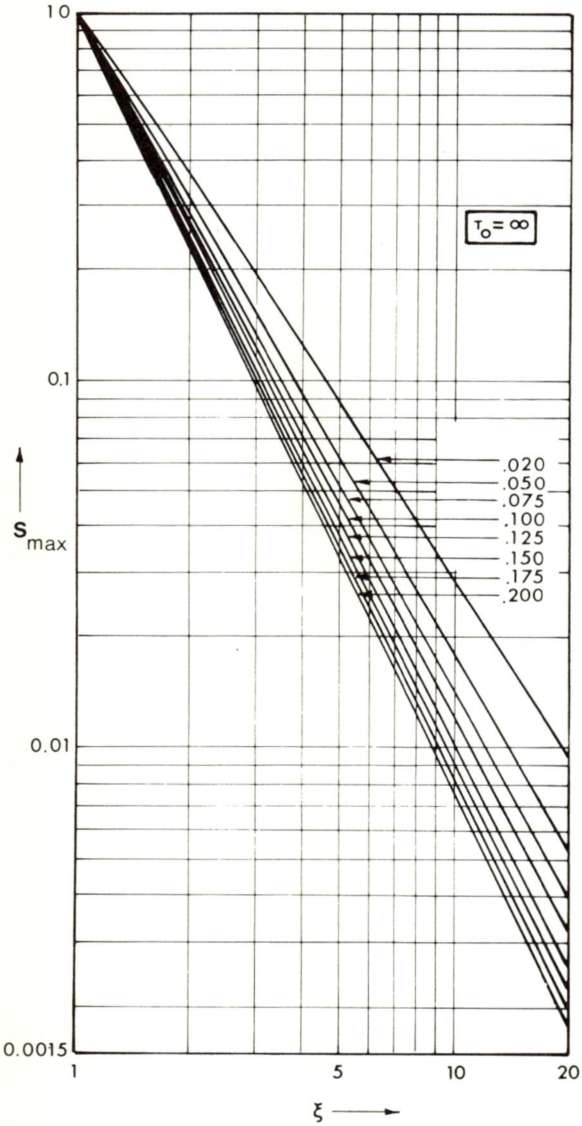

Fig. 51. $S_{r,\max}$ as a function of distance ξ with κ as a parameter ($\tau_0 = \infty$).

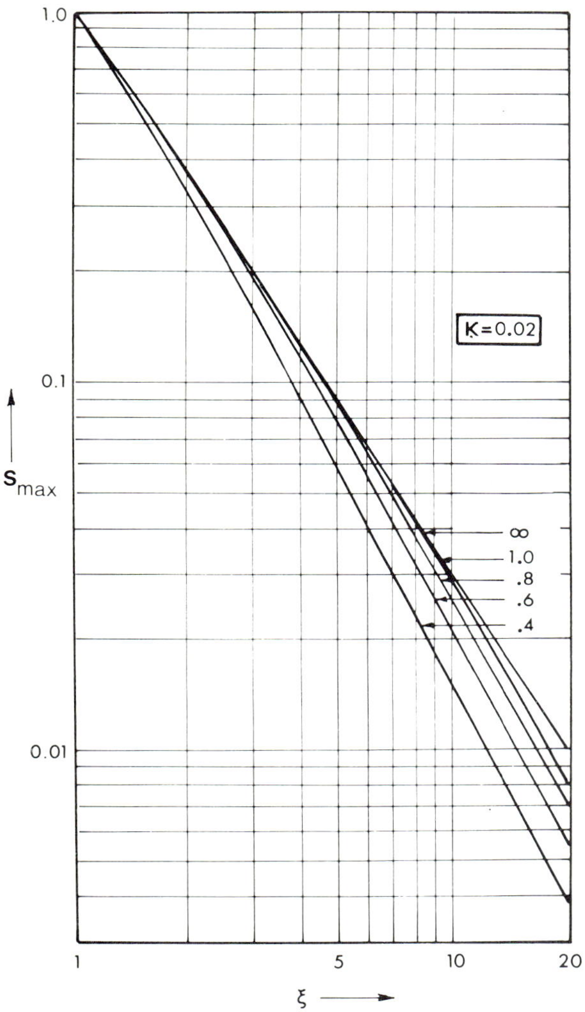

Fig. 52. $S_{r,max}$ as a function of distance ξ with the impulse τ_0 as a parameter ($\kappa = 0.02$).

exhibit dispersion which is caused both geometrically and physically. To bring out these features of the solution, the maximum value, $S_{r,max}$, may be plotted as a function of ξ for different values of κ as shown in Fig. 51. These are the attenuation curves for $S_{r,max}$ based on an input pulse in the form of a Heaviside function. In the doubly logarithmic scale they seem to be straight lines, but a closer examination shows that they are slightly curved.

The preceding section showed, however, that the impulse of the input pulse is a parameter that will affect the attenuation curves. This is exhibited in Fig. 52 for κ =

0.02. One will therefore have a set of curves for each value of κ, and the attenuation curves are thus rather complicated because they depend on two parameters (κ and τ_0).

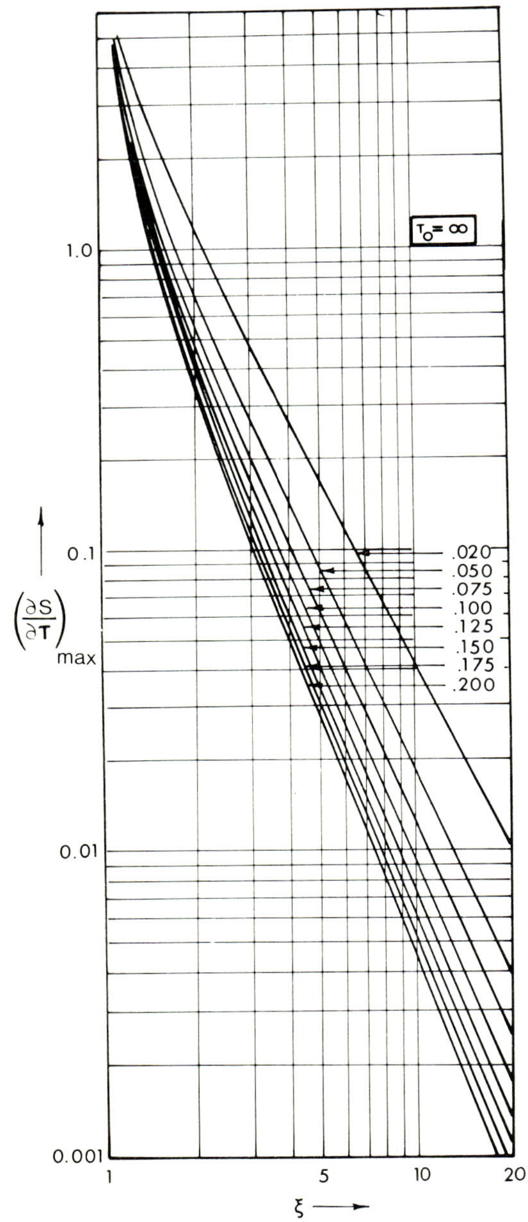

Fig. 53. $(\partial S_r/\partial \tau)_{max}$ as a function of ξ with κ as a parameter ($\tau_0 = \infty$).

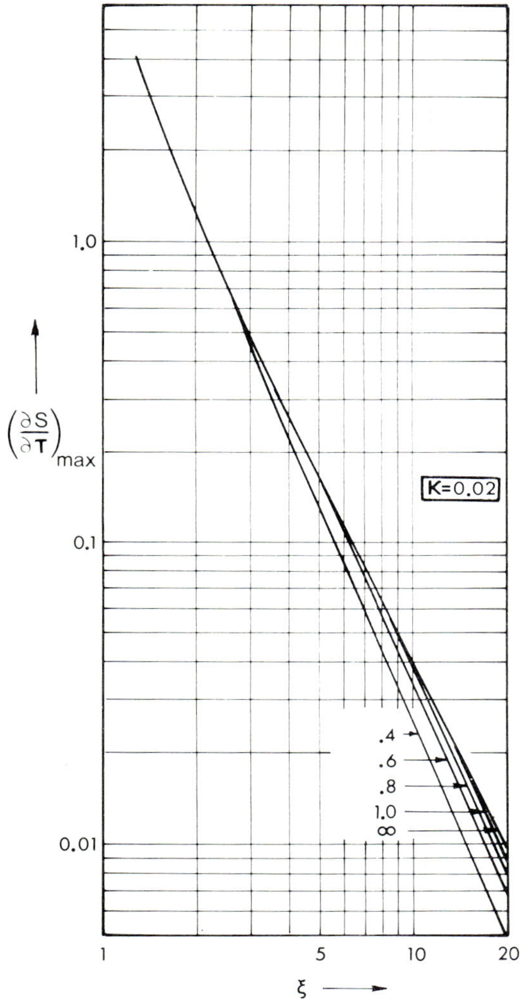

Fig. 54. $(\partial S_r/\partial \tau)_{max}$ as a function of ξ with τ_0 as a parameter (κ = 0.02).

The dispersion of the stress wave is not easily exhibited because it involves many factors. It is, however, possible to use the rate at which the signal initially increases with time as a measure for this feature. The dispersion may thus be measured by $(\partial S/\partial \tau)_{max}$, i.e. the value of the slope of the signal at its first inflection point. If this is done for different values of κ and τ_0 one will obtain the attenuation curves in Figs. 53 and 54. It is seen that, as $\xi \to 1$, the values tend to infinity. This is of course due to the fact that all curves here are computed on the basis of a rectangular input pulse. The values of τ_0 should thus be conceived of as a non-dimensional expression for the impulse of the input pulse. Because of the importance of these curves, the numerical values computed are given in tables and diagrams in Appendixes II and III.

A final remark should be made at this point. In the description of the plane wave (Chapter 2), it was pointed out that a linear relation exists between the stress (σ_x) and the particle velocity ($\partial u/\partial t$; see eq. 2.49). If one for a moment imagines that this relationship could be applied locally also in the present case, the attenuation curves for $S_{r,\max}$ and $(\partial S_r/\partial \tau)_{\max}$ could be conceived of as attenuation curves for particle velocities ($\partial u/\partial t$) and particle accelerations ($\partial^2 u/\partial t^2$), respectively (see further remarks on this in Chapter 11).

THE "SIGNAL" VELOCITY

There is a striking difference between the propagation of stress waves in an elastic and in a Kelvin medium in the sense that no definite propagation velocity can be deduced from the differential equations in the latter case corresponding to c or c_t in the former. If, however, attention is fixed on the maximum value $S_{r,\max}$ occurring at a distance ξ and time τ at which this value occurs, one will find that a plot of ξ versus τ gives an almost straight line as shown in Fig. 55. The slope of this

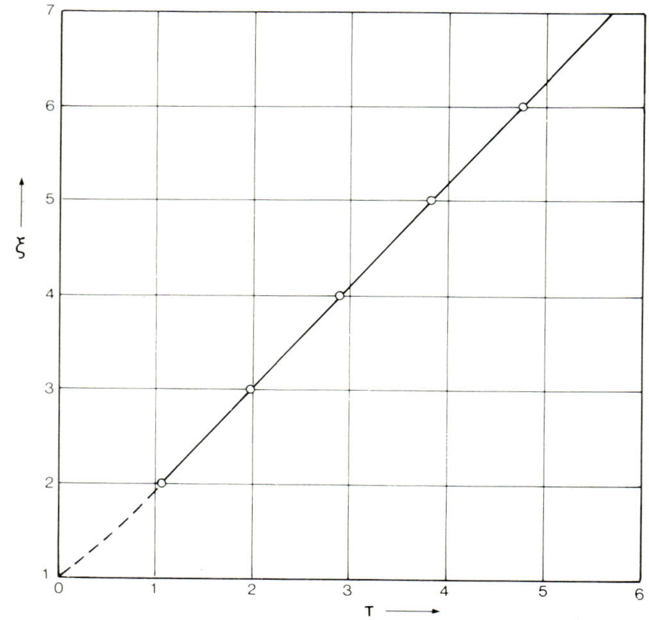

Fig. 55. Position (ξ) of the maximum amplitude of the stress wave as a function of τ (spherical wave, $\tau_0 = \infty$, $\kappa = 0.125$).

line is the velocity with which the maximum amplitude $S_{r,\max}$ would be propagated and may be termed the "signal" velocity. This velocity will depend on τ_0 as well as κ. However, within the range of variation of κ and τ_0 in the tables (Appendixes II and III), this velocity (c_s) may be expressed by the approximate formulae. For $\tau_0 = \infty$:

$$\frac{c_s}{\sqrt{E_1/\rho}} = 1 + 0.038 \left(\sqrt{1 + \frac{\kappa}{0.036}} - 1\right) \qquad [7.22]$$

For $\tau_0 = 1.0$:

$$\frac{c_s}{\sqrt{E_1/\rho}} = 1 + 0.012 \left(\sqrt{1 + \frac{\kappa}{0.012}} - 1\right) \qquad [7.23]$$

Within an accuracy of 10% the following approximation is good enough:

$$c_s = \sqrt{\frac{E_1}{\rho}} \qquad [7.24]$$

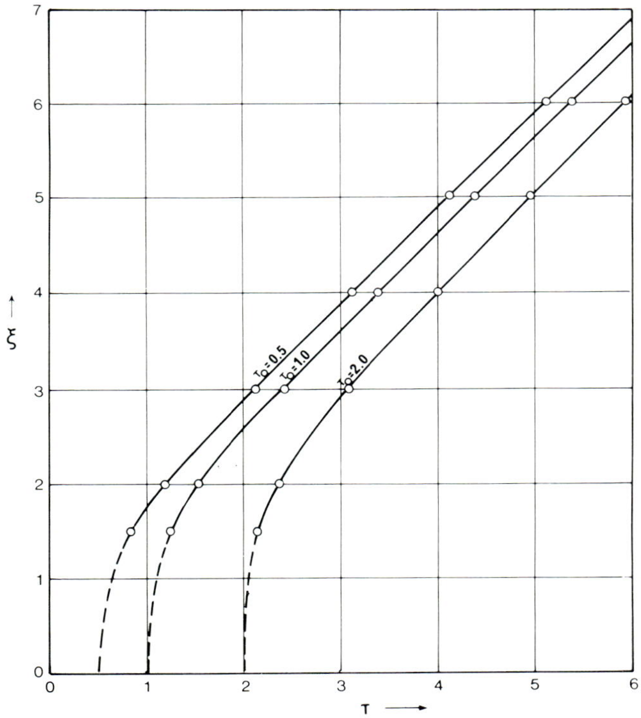

Fig. 56. Position (ξ) of the maximum amplitude of the stress wave as a function of τ (plane wave).

In Fig. 56 the same plot has been made for the plane wave from Chapter 6 as in Fig. 55 for the spherical wave. Again it is observed that the "signal" velocity (expressed as the slope of the curves) approaches a finite value for larger values of ξ. This velocity is seen to be very well expressed by eq. 7.24.

Chapter 8

EXPERIMENTAL STUDY OF STRESS WAVES IN ROCKS

GENERAL REMARKS

Stress waves in rocks may be created in many ways. Landslides, volcanic action, earthquakes and explosions are a few examples. The situations in which stress waves in rocks play an important role are even more numerous. These different situations have been made the object of study and large fields of research such as seismology have been established. Because of the complexity of the situations, in which the study of stress waves in rocks becomes important, it is necessary to limit the scope of the present text to keep it within a managable size.

The increasing trend to use rock massives for the construction of underground facilities like shelters, storage rooms, power stations, etc., is mainly due to our increasing efficiency in the excavating technology. In this context it becomes imperative to study the stress waves created by explosions in the immediate neighborhood of the detonating charge, the propagation of such stress waves and their effect on underground cavities. This is a field where until recently very few experimental results were available, and those which were showed a great scatter which made an attempted theoretical approach to the problem seem rather unrewarding. The remainder of this text represents an attempt to build up a theoretical foundation for the study of this type of problems.

EXPERIMENTAL LAYOUT

The first and most important thing to realize when stress waves in rocks are to be studied experimentally is the fact that such studies cannot be made on rock samples in the laboratory. Such samples are not representative of the rock massive itself with its cracks, layers and other discontinuities. Thus, significant measurements will have to be made over distances large enough to encompass a portion of the rock massive which is representative of the entire massive's properties as a transmission medium. The smallest size of an experimental layout in which this condition will be fulfilled is a matter for experimental investigations to decide.

The scatter in experimental data obtained from measurements of stress waves in rocks is to some extent inevitable. However, part of this scatter may be recognized as systematic errors, and the source of such systematic errors may sometimes be identified. Thus, it becomes possible to correct for them, but this implies that specific measures must be taken already in the experimental layout. It is thus seen that the experimental layout must be carefully planned if sensible conclusions are to be hoped for.

Because the aim is to study the stress waves created by detonating charges, it is natural to also use detonating charges in the experimental investigations. It will, however, become evident later, that only under this condition will the results be applicable for certain purposes.

It is not the intention of the present text to go into a detailed study of the instrumentation of the experiments. A working knowledge of how the experiments are made is, however, crucial for the understanding of the data processing. Assume now the simplest possible experimental layout as sketched in Fig. 57 where two holes are drilled into the rock. In hole A a charge is placed and in hole 1 a pickup is fixed containing a measuring device which either may be a strain gage measuring (presumably) the local strain as a function of time or an accelerometer measuring (presumably) the local particle acceleration as a function of time. The charge should be confined in order to obtain a spherical wave. This may be achieved by pouring concrete in hole A. Sometimes even water will give an adequate confinement of the charge. The pickup will upon detonation of the charge register the signal due to the spherical wave but it will (somewhat later) also register the wave reflected from the surface. The depth of the holes must therefore be made large enough, so that the signal from the reflected wave arrives after the main portion of the primary signal from the spherical wave has been recorded. The depth thus depends on the distance, d, over which the wave travels before it is recorded, and this length d must be made great enough to meet the requirements stated earlier.

Because it is desired to monitor the wave as it travels, more pickups may be placed in additional holes at given distances from the charge. One arrives in this way to an experimental layout such as sketched in Fig. 58. The holes A, B, ... H are

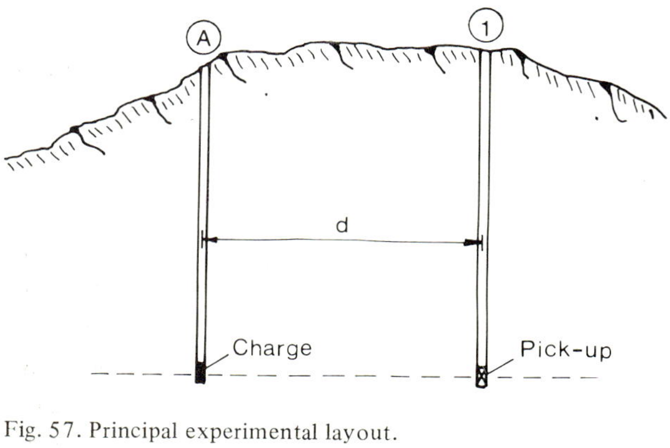

Fig. 57. Principal experimental layout.

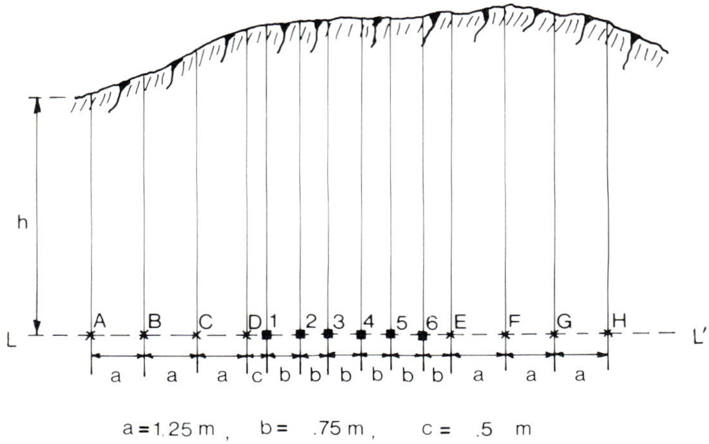

Fig. 58. Standard experimental layout.

used for the charges and in the holes 1, 2, ... 6 the pickups are placed. By detonating in all eight identical charges and recording each time at six pickups one obtains 48 signals, measured at various distances, from which an experimentally obtained attenuation curve may be deduced. The result of such experiments will be discussed below, but before this can be done several other features of the experiment must be commented upon.

SOURCES OF SYSTEMATICAL ERRORS

It was previously mentioned that experimental data obtained in rocks in the present context exhibit a great scatter which to some extent is caused by sources creating systematical errors and to some extent must be considered inevitable. To identify sources of systematical errors thus becomes very important in an effort to improve the reliability of experimental data.

Considering the data one obtains in a standard layout as shown in Fig. 58, one may ask the following question: Will two identical charges when detonated in two seemingly identical holes under identical confinement necessarily create identical peak pressures (denoted by p_0 in the theoretical approach) at the location of the charge? The answer is no. Local differences in the rock material immediately adjacent to the charge in the two seemingly identical holes or the combustion process during detonation of the charge being more or less complete could account for differences in the created peak pressures. Because it is believed that a linear theory is adequate for the description of the wave propagation, the introduction of a "*charge factor*" presents itself as an obvious way of correcting for this systematical error (see section *Data evaluation for a standard experiment*).

The next question one might ask is the following: Will the signals from two identical pickups at seemingly equal distances from one single detonating charge necessarily be identical? Again the answer is no. There are several practical reasons for this answer. First, it is realized that holes, which are supposedly drilled vertically, have a tendency always to become somewhat slanted. Even though corrections are usually made to adjust for this, the determination of a pickup's exact position is not accurate. Secondly, the pickups have a directional sensitivity. Even though precautions are usually taken to insure an exact alignment along the direction of the propagation of the wave, the operation of fixing the pickup's direction is again a source of inaccuracy. Finally the pickups are to be cemented into their holes to make them react with the surrounding rock. The process of cementing these pickups in the hole is always difficult, and on occasion imperfections occur accounting again for systematical errors in the pickup signal. In all these cases, however, it can be shown that a "pickup factor" can be introduced to eliminate the influence of the systematical error. A detailed account of how this is done, and under what conditions it is permissible to do so, will be given when dealing with the data evaluation.

This short exposition of some of the sources of systematical errors in the data has hopefully demonstrated the necessity of a detailed scrutiny of the data before drawing any conclusion from them. Very often this scrutiny will not consist of an inspection of the data themselves only, (these being maximum amplitudes for instance) but the whole signal from each pickup must usually be inspected. Thus, the evaluation of the experimentally obtained information cannot generally be done by indiscriminately applying available data processing techniques.

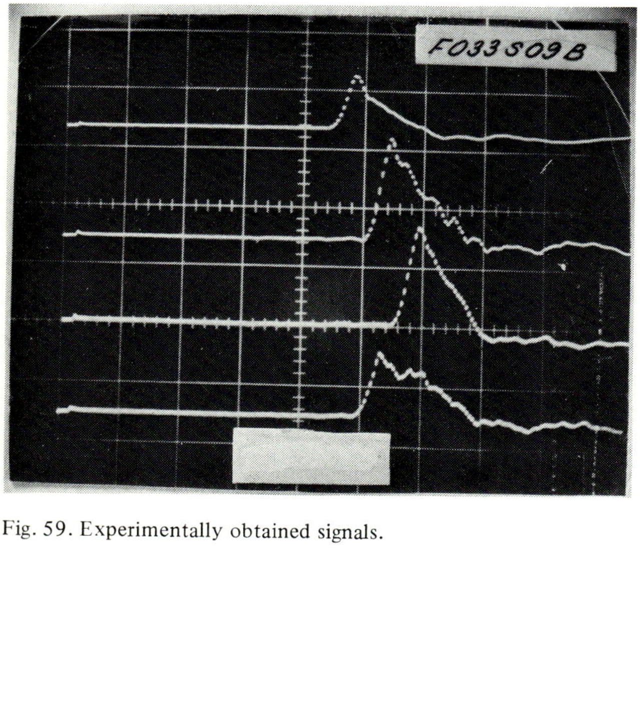

Fig. 59. Experimentally obtained signals.

DATA EVALUATION FOR A STANDARD EXPERIMENT (F33)

When a standard experiment has been performed (in a layout given in Fig. 58) the results will come out as recorded signals from each pickup. Examples of such signals are shown in Fig. 59 from an experiment where the pickups were measuring local strain with strain gages as sensors. The maximum amplitudes from the different pickups are seen to be displaced in time (along the horizontal axis) and because the position of each pickup is known, the time interval between the occurrence of the maximum amplitudes gives the necessary information for the "signal velocity", c_s, to be determined.

From each signal it is now possible to pick out the maximum amplitude, A_i, which in the present case is measured in μ-strain, i.e. a dimensionless quantity which is related to the stress by the calibration of the pickups.

Data evaluation is now best shown through the results from an actual experiment, F33. In this case an experimental layout was used which was a slight modification of the standard layout in Fig. 58. Only five pickups were used, and the spacing between them was $b = 0.75$ m. The layout was symmetrically designed around the middle pickup and the spacing between the holes for the charges was $d = 1.25$ m. The minimum depth was chosen as $H = 6.00$ m to avoid interference from reflected waves. When the holes were drilled, their actual position along line L–L′ in Fig. 58 was measured. The results of the experiment can then be recorded as a number of coordinates (d_i, A_i, where d_i represents the distance from the charge and A_i the recorded maximum amplitude measured at that distance). These data are given in Table III and they are plotted in Fig. 60 in a doubly logarithmic scale. The first thing one realizes is that the data from the pickups closer to the charge than 2.0 m may be unreliable due to errors in the alignment along line L–L′. Thus they are not further considered. (Additional reasons for this will be given in Chapter 9.)

The next step is to realize that the plotted data in Fig. 60 should exhibit a linear relationship, i.e. the relationship is expressed as a straight line:

$$\ln A_i = \lambda \ln d_i + b \qquad [8.1]$$

which would be equivalent to an experimentally established attenuation curve. (This is strongly supported by the theoretical results in Chapter 7.) Introducing the notations:

$$x_i = \ln d_i, \quad y_i = \ln A_i \qquad [8.2]$$

the straight line would be:

$$y = \lambda x + b \qquad [8.3]$$

and the deviation, v_i from this could be expressed for each set of data as:

$$v_i = y_i - \lambda x_i - b \qquad [8.4]$$

TABLE III

A_i data from experiment F33

Charge in	Pickup	d_i (m)	A_i (μ-str)	Charge factor β	A_i (corr.)	Pickup factor
A	1	5.08	168		228	
	2	5.83	147		187	0.94
	3	6.59	114	1.35527	155	
	4	7.22	76		103	
	5	8.00	59		94	1.17
B	1	3.82	356		382	
	2	4.57	257		259	0.94
	3	5.33	–	1.07358	–	
	4	5.96	150		161	
	5	6.74	104		131	1.17
C	1	2.63	–		–	
	2	3.38	590		390	0.94
	3	4.14	509	0.70398	358	
	4	4.77	316		222	
	5	5.55	272		224	1.17
D	1	(1.24)	(1510)		(1810)	
	2	(1.99)	(980)		(1104)	0.94
	3	2.75	601	1.19842	720	
	4	3.38	377		452	
	5	4.16	210		294	1.17
E	1	3.72	354		425	
	2	2.97	522		589	0.94
	3	2.31	731	1.20081	878	
	4	(1.58)	(988)		(1186)	
	5	(0.80)	–		–	1.17
F	1	4.84	220		275	
	2	4.09	311		366	0.94
	3	3.33	404	1.25107	505	
	4	2.70	510		638	
	5	(1.92)	(900)		(1317)	1.17
G	1	6.20	140		135	
	2	5.45	218		198	0.94
	3	4.69	256	0.96657	247	
	4	4.06	330		319	
	5	3.28	470		532	1.17
H	1	7.51	180		101	
	2	6.76	307		161	0.94
	3	6.00	319	0.56046	179	
	4	5.37	–		–	
	5	4.59	379		249	1.17

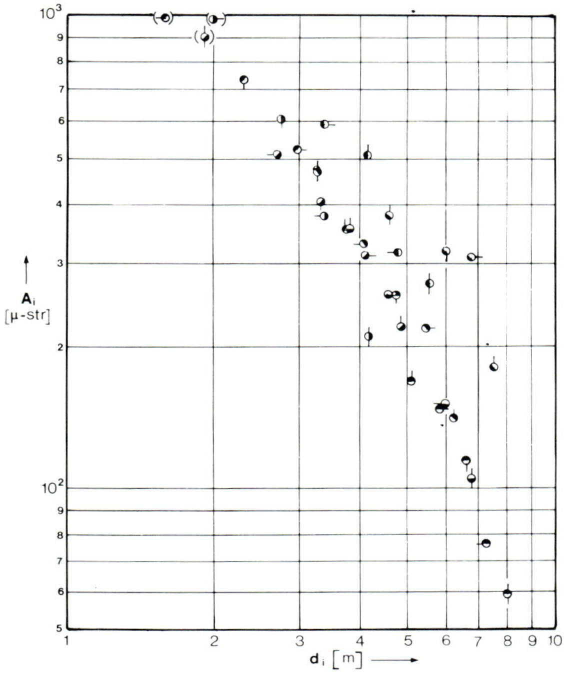

Fig. 60. Recorded data from standard experiment F33.

Now attention is drawn to the fact that, even though the charges detonated in holes A, B, ... H are of equal weight (W), the necessity of introducing "charge factors" (β_A, β_B, ... β_H) for each charge is present. The effect of the charge factors is to adjust for the difference in the input impulse, and in the diagram in Fig. 60 this effect will show up as a vertical displacement of the data points by the same amount for each point from one shot. The charge factor will of course vary from shot to shot. Thus, the values of A_i should actually be replaced by $\beta_A A_i$, $\beta_B A_i$, etc., and 8.4 should, therefore, be replaced by:

$$v_i^{(A)} = y_i + \ln \beta_A - \lambda x_i - b$$

$$v_i^{(B)} = y_i + \ln \beta_B - \lambda x_i - b \qquad [8.5]$$

etc.

The coefficients λ and b of the straight line, as well as the "charge factors" β_A, β_B, ..., are now determined by linear regression, i.e. the method of least squares. The sum of the squares of all v_i is to be a minimum. Thus:

$$V = \sum_A (v_i^{(A)})^2 + \sum_B (v_i^{(B)})^2 + \ldots \qquad [8.6]$$

whereupon the following conditions give the equations to determine the unknown quantities:

$$\frac{\partial V}{\partial k} = 0, \quad \frac{\partial V}{\partial b} = 0, \quad \frac{\partial V}{\partial(\ln \beta_A)} = 0, \quad \frac{\partial V}{\partial(\ln \beta_B)} = 0, \ldots \qquad [8.7]$$

If this is carried out in detail it will be found that one arrives at an indeterminate set of linear equations for the unknown quantities. This reflects the fact that all data might be multiplied by a common factor which in Fig. 60 would correspond to a vertical displacement of all points by the same amount. Because no single "shot" in the series has any preference over the others, it is fair to seek the "mean vertical position" of all "shots", which applied to the charge factors implies:

$$\ln \beta_A + \ln \beta_B + \ldots + \ln \beta_H = 0 \qquad [8.8]$$

In this way a complete set of equations for the determination of the unknowns is established.

The data of F33 were subjected to the evaluation procedure outlined above and the result is shown in Fig. 61. (It is important in the evaluation procedure that pickups and shots can easily be recognized, therefore, the legend given in Table VIII (p. 142) is used throughout this text.) In Fig. 61 it is noticed that all data from pickup no. 5 appear under the straight line, and that these seem to gather along a line which would be parallel to the one shown, but displaced somewhat downwards. The same applies to data from pickup no. 2. This indicates that both of them display a systematic deviation. (This conclusion rests on the fact that the stress waves created travel in both directions.) One might try to compensate for this directly by introducing a "pickup factor", but it is preferred to do the evaluation over again, leaving out the data from these two pickups. One then arrives at the following values of λ and b:

$$\lambda = -1.7971, \quad b = 8.32158 \qquad [8.9]$$

The obtained "charge factors" β are listed in Table III.

The deviations from this line exhibited by the data from the pickups 2 and 5 are then corrected for by "pickup factors" α_i which are determined separately for each pickup such that the mean deviation is zero. This gives the "pickup factors".

$$\alpha_2 = 0.94, \quad \alpha_5 = 1.17 \qquad [8.10]$$

which show the systematic deviation to be 6% and 17%, respectively for the two pickups. One may then correct all data according to these results, and these are again given in Table III. The linear regression process for these corrected data gives:

$$\lambda = -1.7812, \quad b = 8.30187 \tag{8.11}$$

and a standard deviation:

$$m = \pm\ 10.86\% \tag{8.12}$$

The final result is shown in Fig. 62.

A final remark should be added to these results. The fact that the deviations in the results from the pickups 2 and 5 are really systematic shows up in different ways.

(1) The deviations must be of the same order of magnitude irrespective of the way in which the wave travels. (Note: If the data are obtained with only one direction of travel no justification can be advanced for assuming a systematical

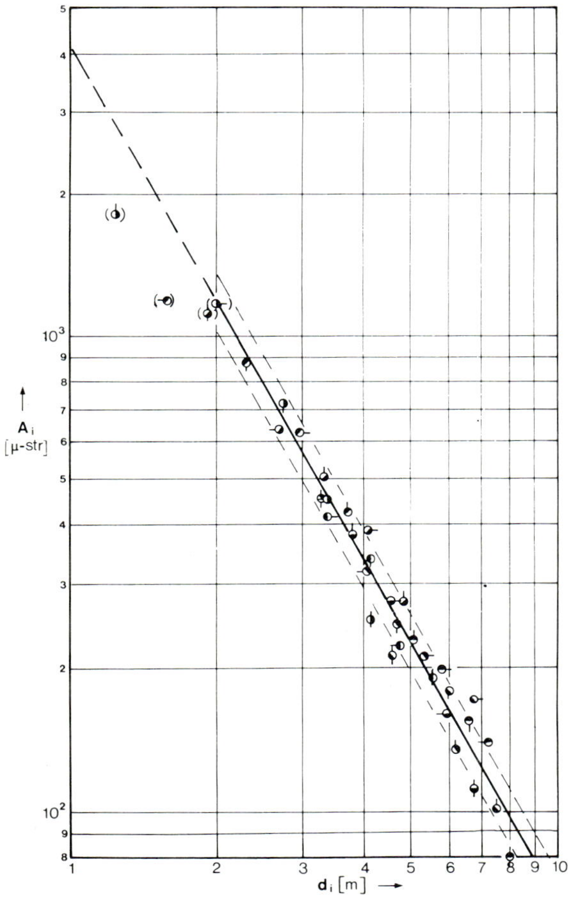

Fig. 61. Data of F33 corrected by the charge factors and compared with the straight line determined by the evaluation.

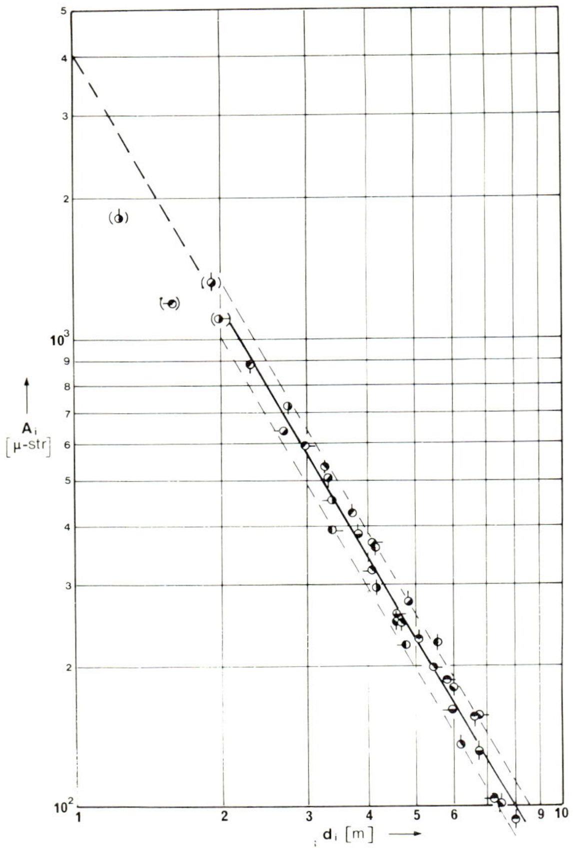

Fig. 62. Result of experiment F33 after evaluation.

deviation. The data may then express a physical reality like a local inhomogeneity in the rock.)

(2) The results of the data evaluation must be very little affected by whether or not the data suspected of exhibiting systematic deviations are taken into account. In the present case compare 8.9 with 8.11.

(3) The signals from which the data are obtained must show no sign of "strange" deviations when compared with similar signals from other pickups.

UNIQUENESS OF A STANDARD EXPERIMENT

The experiment described in the preceding section will not have any significance for practical purposes unless it can be shown to give information of fundamental

TABLE IV

A_i data from experiment F34

Charge in	Pickup (i)	d_i	A_i	Charge factor β	A_i(corr.)
A	1	4.95	240		242
	2	5.75	170		172
	3	6.70	125	1.00912	126
	4	7.27	103		104
	5	8.09	79		80
B	1	3.75	374		369
	2	4.55	243		240
	3	5.50	186	0.98789	184
	4	6.07	166		164
	5	6.89	112		111
E	1	3.75	465		394
	2	2.95	600		509
	3	2.00	1360	0.84837	1154
	4	–	–		–
	5	–	–		–
F	1	5.09	192		227
	2	4.29	229		271
	3	3.34	405	1.18238	479
	4	2.77	474		560
	5	–	–		–

character about the propagation of stress waves in rocks in general. How it is possible to assign such a significance to the results will be exhibited in the next chapters. Before this is done, one must however ascertain that the experiment is unique in the sense that when it is repeated in the same rock massive but on another location within this massive, the results will be the same. This can only be done by repeated experiments and to lay the foundation for future reference, two more experiments, F34 and F33b, made in the same rock massive as F33, will be reported here.

The data from F34, obtained with the same charges (measured in weight of the same explosive) as in F33, are given in Table IV. The evaluation procedure is the same as before, but this time no systematic deviation in the pickup readings could be found. However, only charges in holes A, B, E and F gave reliable results which accounts for the fewer data in this experiment. The raw data (d_i, A_i) are plotted in Fig. 63 to give an impression of the apparent scatter.

The evaluation procedure given in the preceding section will in the present case render the charge factors β in Table IV and the following constants λ and b for the straight line:

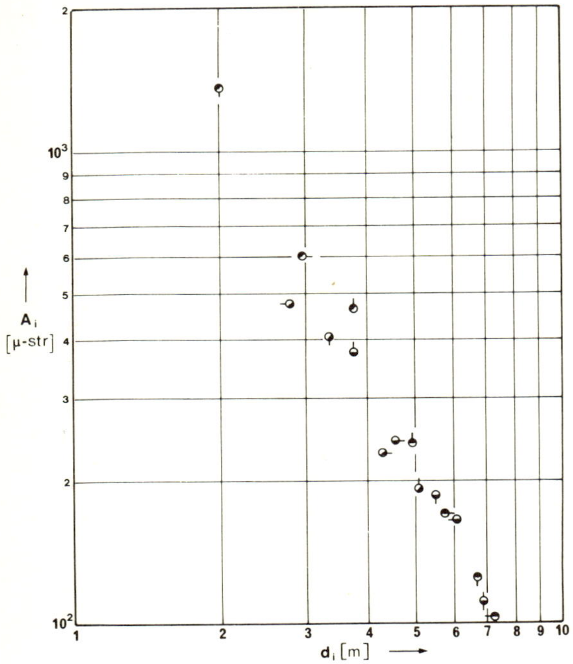

Fig. 63. Raw data from experiment F34.

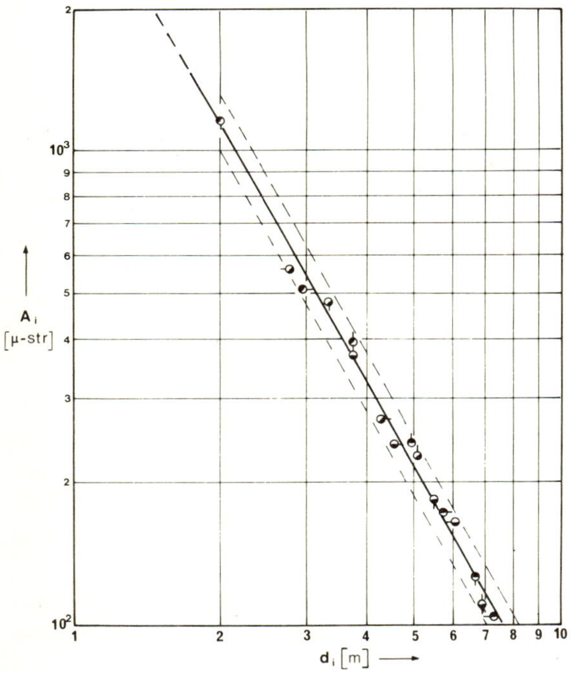

Fig. 64. Final result of the data evaluation for experiment F34.

$$\lambda = -1.8250, \quad b = 8.30359 \qquad [8.13]$$

and a standard deviation:

$$m = \pm 7.43\% \qquad [8.14]$$

The final result is exhibited in Fig. 64.

This result has to be confronted with the results from F33. This is done in Fig. 65 where the two straight lines have been plotted in one diagram. It is seen that well within the accuracy of the measurements the two straight lines do indeed give the same answer. However, it must be remembered that these two experiments were done with the same charge in all cases. This charge was $W = 0.5$ kg Pentritt-wax, an explosive with a high combustion rate. The results for other values of W cannot be predicted from the results obtained. The standard experiment was therefore repeat-

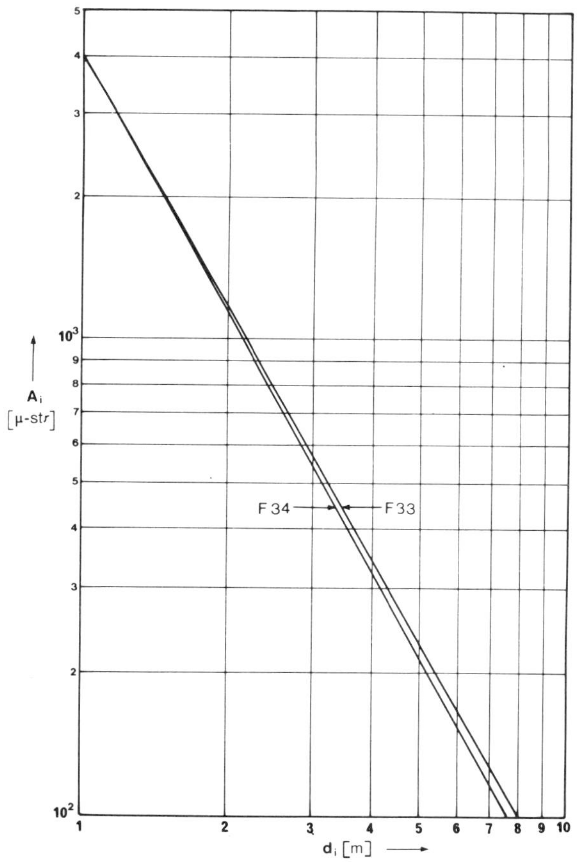

Fig. 65. Comparison of the results from F33 and F34.

ed with charges W = 2.5 kg in the same holes as those used for F33. The detailed data and their evaluation for this experiment, F33b, are omitted here but the constants λ and b for the straight line were determined as:

$$\lambda = -1.7222, \quad b = 8.80030 \tag{8.15}$$

These values are close to the ones obtained in 8.11 and 8.13 as far as λ is concerned, but the value of b is much different. This brings the exposition of the experimental investigation of stress waves in rocks to the point where theoretical results must be drawn upon to give a more complete understanding of what happens. (The final resolution of this problem is given in Chapter 11.)

Chapter 9

EXPERIMENTAL RESULTS COMPARED WITH THEORY

INTRODUCTORY REMARKS

In the first chapters of this text emphasis was placed on an exposition of theoretical approaches to the propagation of stress waves in solids. In Chapter 8 a straightforward experimental approach to the study of stress waves in rocks was given. So far, however, no direct connection between these two approaches has been shown, although a few references to theoretical results were made when discussing different aspects of the experiments. The scope of the present chapter is to exhibit the applicability of the theory for a Kelvin medium to the propagation of stress waves in rocks. It should, however, be stressed already at this point that this does not imply the assumption that the rock is a visco-elastic medium. This point will be elaborated on at a later stage.

GENERAL FEATURES OF THE THEORETICAL RESULTS

A general feature of the theoretical solutions is that they are based on a linear theory. The solutions give the non-dimensional stress as a function of a non-dimensional distance and a non-dimensional time. The problem is in other words assumed to contain a definite distance (a), a definite velocity (c) and a definite stress (p_0) with which the above-mentioned quantities can be made dimensionless.

It has been shown that the theory for a spherical wave does indeed exhibit a finite velocity c_s which corresponds to the signal velocity one can measure. However, the finite length, a, which in the theory appears as the radius of a spherical cavity inside which the pressure, p, acts to create the stress wave, does not have an immediate counterpart in the experimental case. Nor is the peak value p_0 of this pressure an immediately measurable quantity. And lastly, the assumption that the physical reality of stress wave propagation in rock may be adequately expressed by a linear theory must be made plausible before it can be accepted.

The attention is first drawn to the attenuation curves which show the attenuation of the maximum amplitudes of the stress wave with increasing distance from the origin. When plotted in a doubly logarithmic scale, these curves turn out to become straight lines at sufficiently large distances from the origin. This holds true for the experimental results as well as for the theoretical results. The slope, λ, of these lines thus becomes a significant parameter of the stress-wave propagation. Now the results from the different cases worked out for an elastic medium (Chapter 3) show that λ in all cases approaches the value -1. This is in contradiction with experimental evidence which thus rules out the possibility of using the theory for stress waves in an elastic medium to describe the propagation of stress waves in rocks.

In Chapter 7 the stress wave propagation in a Kelvin medium was studied and the results showed that the attenuation curves could be considered as almost straight lines in the distance range $10 \leqslant \xi \leqslant 20$. The slope, λ, of these lines does not only depend on the value of damping parameter κ but also on the impulse of the input pulse expressed by τ_0. The rather complicated dependence of λ on κ and τ_0 must, therefore, be studied in detail and the result must be confronted with experimental evidence before judgement can be passed on the applicability of the theory.

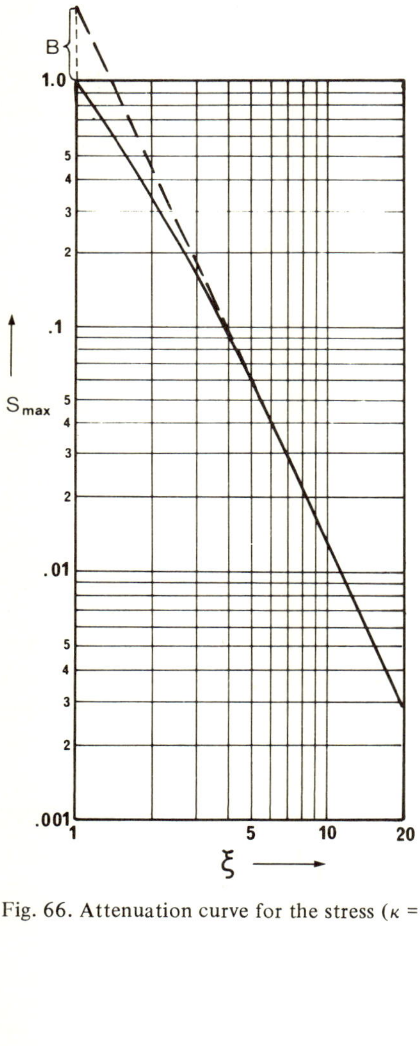

Fig. 66. Attenuation curve for the stress (κ = 0.05, τ_0 = 0.8).

Fig. 66 shows the theoretical attenuation curve for $\kappa = 0.05$ and $\tau_0 = 0.8$. It is seen that its main trend is given by a straight line for $\xi > 10$. This straight line may be used as an approximation to the theoretical curve for $\xi > 5$ within the accuracy of the experiments. If, however, the straight line is continued all the way to $\xi = 1$, the value will be B. The value of B determines the vertical position of the straight line and thus the attenuation will be completely determined through the values of B and λ. Fig. 67 shows a plot of all the straight line approximations to the attenuation curves for values of κ and τ_0 given in the tables Appendixes II and III. The lines are shifted by one decade for each value of τ_0. It is brought out that not only are the slopes (λ) of the lines determined by the values of parameters κ and τ_0, but the vertical position of these lines is also determined by the same parameters. For a true statement on the attenuation it is just as important to know the vertical position of the line as to know its slope. This aspect of the attenuation will be discussed in detail at a later stage. For the moment the general trend of the theoretical solution will be considered.

The aim is to be able to apply the theoretical results to experimental data. These are limited in their accuracy, and any approximation to the theoretical results which is well within this accuracy will therefore be sufficient. Within the range of

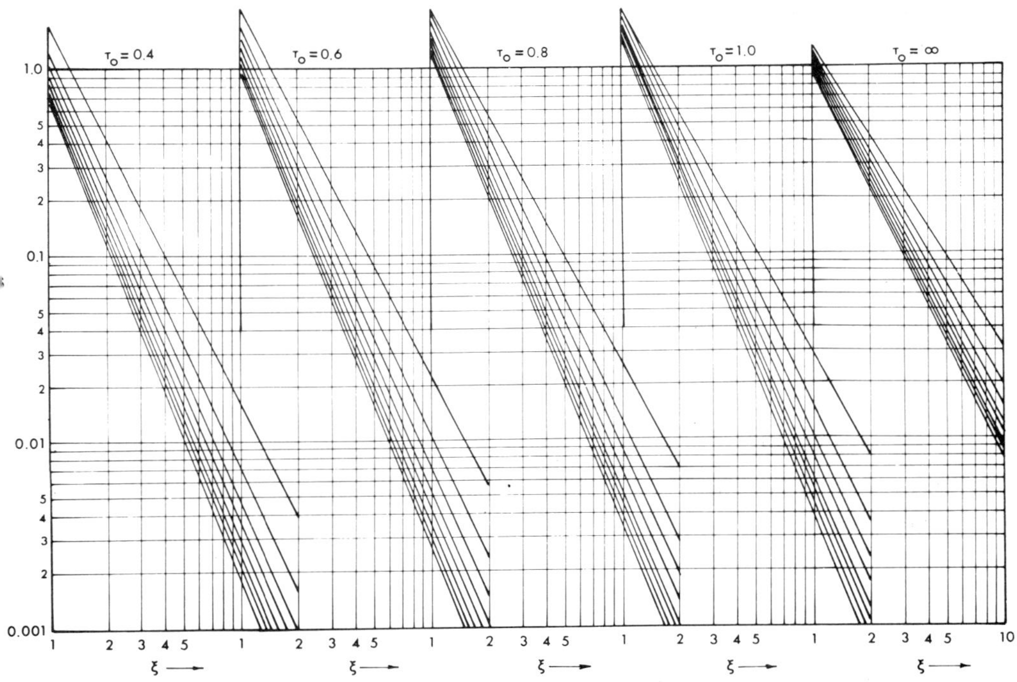

Fig. 67. Display of attenuation curve approximations for different values of κ and τ_0.

values of κ and τ_0 for which tables exist for the theoretical solution, the following expressions for the slope λ are valid within an accuracy of 1% within the range $10 \leqslant \xi \leqslant 20$:

$$\left.\begin{aligned} \lambda_\infty &= -(1 + 1.5649\, \kappa^{0.22615}) \\ \lambda &= \lambda_\infty - 0.5222\, e^{-\alpha\tau_0} \\ \alpha &= 0.010897\, \kappa^{-1} + 0.09307 \end{aligned}\right| \qquad [9.1]$$

Here λ_∞ represents the slope for $\tau_0 = \infty$, i.e. for the input pulse equal to the Heaviside function.

A few remarks to these formulae seem appropriate. It should be noticed that:

$$\text{when } \kappa \to 0 \quad \text{then } \alpha \to \infty, \quad \lambda = \lambda_\infty = -1 \qquad [9.2]$$

This means that when the elastic case is reached, the slope obtained from these formulae is in accordance with the theoretical results for this case. The validity of the formulae is restricted to the range of distances given above. It is furthermore at this stage convenient to stress the smallest value of ξ for which the straight line has any significance. This is a point which will be considered later.

Having in this way established expressions which adequately give λ as a function of parameters κ and τ_0, the next step is to fix the vertical position of the straight lines through similar formulae for B.

If B_∞ is the value of B for $\tau_0 = \infty$, the dependence of B_∞ on κ may be expressed as follows:

$$B_\infty = 1 + 0.47758 \left(\frac{\kappa}{0.15}\right)^{1/8} \left(1 - \frac{\kappa}{0.15}\right) e^{-0.5\kappa/0.15} \qquad [9.3]$$

Then:

$$\frac{B}{B_\infty} = 1 - \left[1 - \left(\frac{\tau_0}{\tau_*}\right)^2 + 0.75 \frac{\tau_0}{\tau_*} \left(1 - \frac{\tau_0}{\tau_*}\right)\right] e^{-\tau_0/\tau_*} \qquad [9.4]$$

where:

$$\tau_* = 0.93918\, \kappa^{0.29} \qquad [9.5]$$

Again it should be stressed that these formulae are obtained as sufficiently accurate representations in the range $0.02 \leqslant \kappa \leqslant 0.2$, $0.4 < \tau_0 < 1.0$, but they exhibit features which makes them reliable outside this range. First:

$$\text{when } \kappa \to 0 \quad \text{then } \tau_* \to 0, \quad B \to B_\infty \text{ and } B_\infty \to 1 \qquad [9.6]$$

This again is in complete accordance with the results for the elastic medium. Secondly:

$$\text{when } \tau_0 \to \infty \quad \text{then } B \to B_\infty \qquad [9.7]$$

as it should.

Formulae 9.1, 9.3, 9.4 and 9.5 are shown in Figs. 68 and 69 for B and λ,

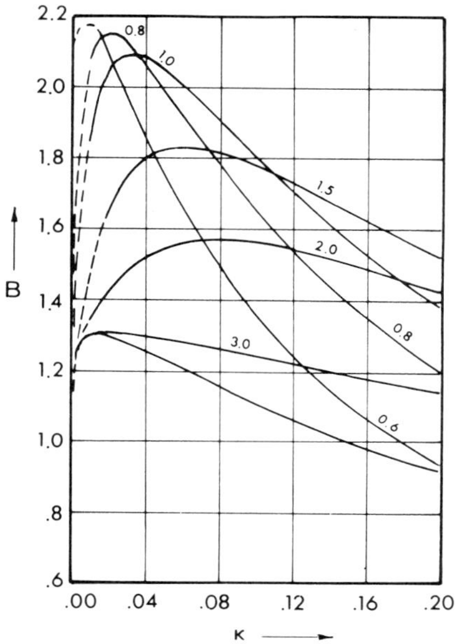

Fig. 68. B plotted as a function of κ with τ_0 as a parameter.

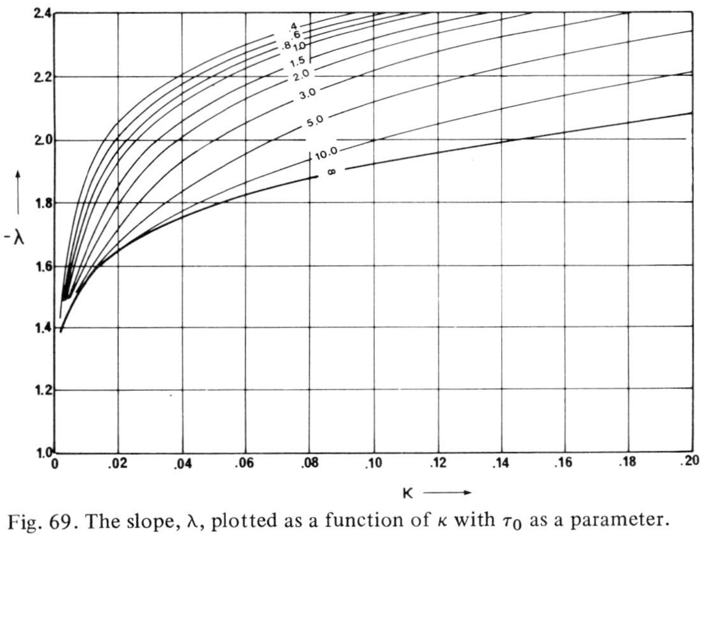

Fig. 69. The slope, λ, plotted as a function of κ with τ_0 as a parameter.

respectively. Some general features of these results should be drawn to attention. In Fig. 68 it is noticed that the curves for $\tau_0 = 5$ and $\tau_0 = \infty$ seem to coincide over a large range of values of κ, thus indicating that all waves with $\tau_0 > 5$ could be conceived of as $\tau_0 = \infty$. This must, however, not be misinterpreted. It should be stressed that the fact that the input pulse is finite will always influence the attenuation curve. The greater τ_0 is, the greater is, however, the distance at which this influence becomes noticeable. It is only because the formulae given here are restricted in their validity to the range $10 \leqslant \xi \leqslant 20$ that one will have this apparent behavior.

The same remarks made to Fig. 68 apply to Fig. 69 although it is noticed that here the value of λ is influenced by τ_0 even when $\tau_0 > 5$.

COMPARISON BETWEEN THEORY AND EXPERIMENT

With the results obtained here and the experimental results exhibited in Chapter 13, the stage has been set for a confrontation between theory and experiment. A comparison between the two does, however, seemingly meet with certain obstacles which may be examplified in the following:

(1) The theoretical attenuation curves appear in terms of non-dimensional coordinates (S, ξ). In contrast, the experimental curves appear in terms of dimensional coordinates (μ-strain and distance measured in meters). The quantities necessary to convert one into the other are the radius (a) of the "sphere" in which the explosion takes place and the peak pressure (p_0) created by the explosion inside this "sphere". In the experiment no such "sphere" exists and measurement of the pressure becomes impractical.

(2) The theoretical results are based on the assumed rectangular input pulse with an impulse given by τ_0. The input pulse in the experimental case remains unknown.

The difficulties pointed out above are crucial for a successful comparison between theory and experiment. The problem of determining the radius a and the impulse τ_0 is a complex one and will be dealt with in detail later. For the present purpose, however, these difficulties are eliminated by the following line of reasoning. The geometry of the charge was as sketched in Fig. 70, in all experiments referred to with 0.5-kg charges. It is not spherically symmetric, but it may be argued that at a radius of 40 cm from its center, the effect of the exploding charge will be almost spherically symmetric. Inspection after detonation of such a charge indicates that the radius of the zone pulverized by the explosion is of the order of 12.5–20 cm. Thus, when the hole is recharged with a charge of 2.5 kg (as in the case F33b) the radius of the sphere at which symmetry will occur may be taken to be the same as before. In the present case then a value of $a = 0.4$ m seems to be an acceptable value for the radius of the imaginary spherical cavity corresponding to the explosion. This

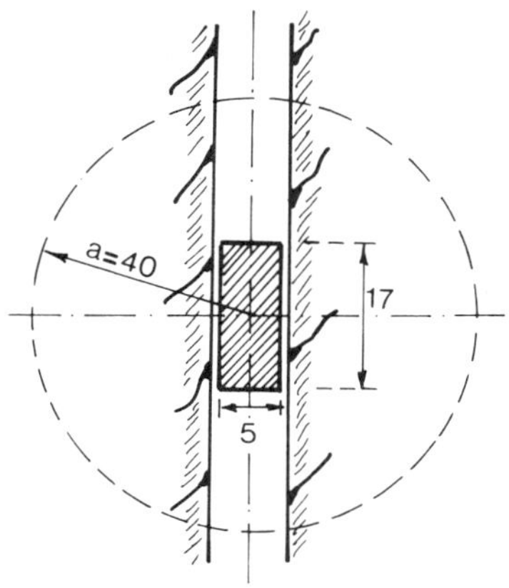

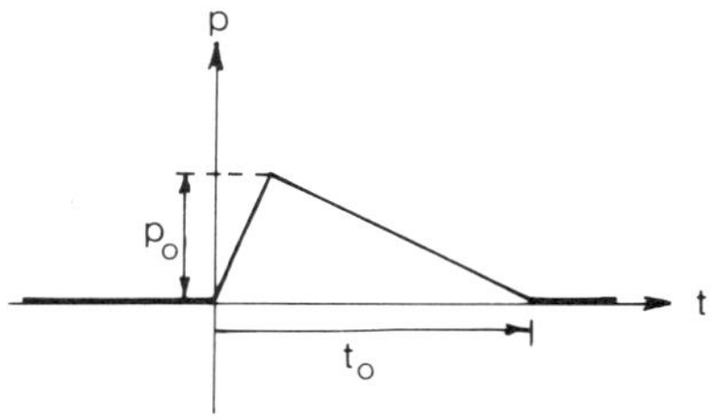

Fig. 70. Dimensions (in cm) of the charge (shaded) placed in the hole.

Fig. 71. The experimental input pulse.

value can then be used to non-dimensionalize the distances measured in meters. It should be mentioned that these considerations will be replaced by a much more rigorous procedure when results from confined explosions are compared with theoretical results (see Chapter 11, section *Complete evaluation of an experiment*). A discussion of the radius a is also given in Chapter 10, section *The radius of the spherical cavity.*

No approximate evaluation of peak pressure p_0 exerted at the above-mentioned sphere can be made directly. However, it will be shown how this can be estimated later by implication.

The fact that rectangular pulses start impulsively, whereas the pressure created by the explosion starts gradually is not a serious discrepancy between theory and experiment. A rectangular input pulse will (in the theoretical case) give a stress-time dependence which, even very close to $\xi = 1$, will exhibit a gradual (but very rapid) increase. This may be conceived of as "the input pulse", and thus an assumed rectangular pulse will only affect the value of a very slightly. Compared to the uncertainty in the estimation of a this discrepancy is negligible.

The parameter τ_0 appearing in the theoretical approach must, however, be found also experimentally. Returning for a moment to Fig. 19 and the exposition of the sinusoidal input pulse in the elastic medium, it is noticed that the value of the duration time, t_0, of the positive phase of the signals does not change considerably over short distances. This is a feature which is preserved in the experimentally observed signals in Fig. 59. Experimental evidence from a series of experiments indicate that t_0 measured as the duration time of the positive phase of the signals close to the explosion may be used to indicate the duration time of the input pulse.

This pulse may be conceived of as a triangular pulse as shown in Fig. 71. The non-dimensional form of this pulse is obtained by normalizing $p(t)$ with the aid of p_0 such that its maximum value becomes equal to 1 and then expressing t_0 in non-dimensional form by means of 7.4:

$$\tau = \frac{t}{a}\sqrt{\frac{E_1}{\rho}} \qquad [9.8]$$

Remembering that τ_0 in the theoretical approach represents the impulse of the non-dimensional input pulse, one obtains in view of its triangular shape:

$$\tau_0 = \frac{t_0}{2a}\sqrt{\frac{E_1}{\rho}} \qquad [9.9]$$

It is thus seen that, in order to make a comparison between theoretical and experimental results, one has to return to the data in the experiments, determine t_0 from the signals and determine $\sqrt{E_1/\rho}$ as the signal velocity (7.24). Then, using the estimated value of a = 0.4 m, one obtains from 9.9 the values of τ_0 for the experiments in question as shown in Table V. In this table also the values of λ and b for the experimentally determined attenuation curves are repeated.

Because the attenuation curves determined both experimentally and theoretically are straight lines in sufficient distance from the charge, the slope of these lines will not be affected by reducing dimensional coordinates to non-dimensional coordinates. The straight lines appear in a doubly logarithmic diagram, and a reduction of distances measured in meters to distances measured with radius a as a unit length will mean only a horizontal displacement of the line parallel to itself. Similarly, a normalizing of μ-strain to a non-dimensional stress would mean a vertical translation of the straight line parallel to itself. The "measured" value of λ is thus to be directly compared with the theoretical value. Using the diagram in Fig. 69, it is seen that with the experimentally determined values of λ and τ_0, it is possible to extract from the diagram the value of κ for each experiment. Fig. 72 shows that part of the diagram in Fig. 69 in which the present data occur. It is observed that each experi-

TABLE V

Results of F33, F34 and F33b

Experiment	τ_0	λ	b
F34	1.051	−1.8250	8.30359
F33	1.256	−1.7812	8.30187
F33b	1.760	−1.7222	8.80030

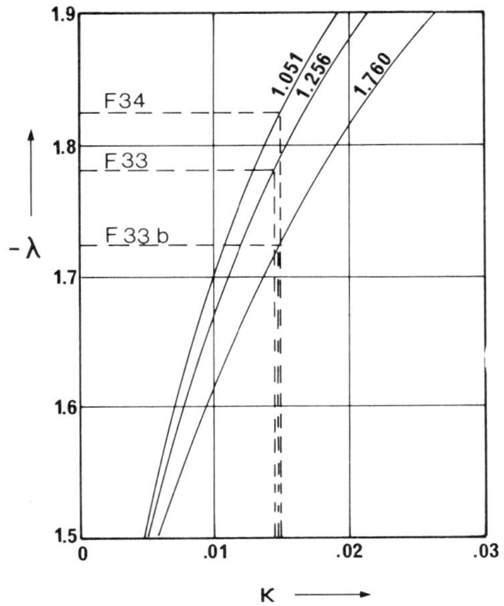

Fig. 72. Enlarged portion of Fig. 69 in which the experimental data of F34, F33 and F33b are plotted.

ment, independently of the others, determines the same value of κ, i.e. κ = 0.015. From this fact the following conclusions may be drawn:

(1) The parameter κ can indeed be conceived of as a true parameter characteristic of the problem.

(2) Whether κ characterizes the material and thus can be conceived of as a true parameter of the material in the thermodynamic sense, cannot be concluded from this. Experiments in the same rock site which exhibits a different "process" must here be considered before any such conclusion can be drawn.

(It should be emphasized at this point that the conclusion drawn does not depend on the results obtained in these three experiments alone. A large number of data from various field experiments points to the same result, but for the presentation here an example has been chosen which shows a remarkable degree of accuracy.)

The next step in the comparison between theory and experiment will be to exhibit the importance of the value of B. If the applicability of the theory is accepted the following line of reasoning holds true:

(1) If the straight lines determined experimentally (attenuation curves) are extrapolated inwards towards the explosion one may find the value A_0^* (in μ-strain) which apparently occurred at a distance a (= 0.4 m) (corresponding to ξ = 1).

(2) Because the values of τ_0 and κ are now known and the theory assumed

TABLE VI

Results of F33, F34 and F33b

Exp.	B/B_∞	B	A_0^*	$A_0 = A_0^*/B$	Charge (kg)
F34	1.08546	1.93722	21500	11098	0.5
F33	1.34144	1.75273	20619	11764	0.5
F33b	1.11438	1.45605	32152	22082	2.5

applicable, the value of B from the theoretical approach gives the factor with which this apparent value at $\xi = 1$ must be reduced to obtain the real value of A_0 at this point.

(3) Because F33 and F34 are identical experiments in one and the same rock massive, the obtained values should be the same (within the accuracy of the experiment).

The computations necessary to show this are found in Table VI.. First it is observed that τ_* and B_∞ with $\kappa = 0.015$ can be determined from 9.5 and 9.3, respectively:

$$\tau_* = 0.27785, \quad B_\infty = 1.30660 \qquad [9.10]$$

(These are the same for all three experiments). Then B is determined from 9.4 for each experiment. From the straight lines (given by λ and b) the values of A_0^* are then computed at the distance 0.4 m, and their real values of A_0 are then easily found. It is seen that the results obtained in F34 and F33 deviate less than 3% from their mean value, a remarkably good agreement. The result obtained in F33b is, however, much higher than the other two. This is of course caused by the fact that the charges in the first two cases were 0.5 kg compared with 2.5 kg in the latter case. This raises certain questions concerning the conversion of values obtained with one charge magnitude to significant values for another charge magnitude, a problem that will be dealt with in Chapter 10. In as far as the importance of B is concerned, the results have shown that one may by means of the theory obtain information about what happens close to the charge, and the usefulness of the theory has again been demonstrated.

DETONATION IN CHAMBERS

One of the most difficult parameters to determine experimentally is the radius (a) of the imaginary spherical cavity with which the real explosion is replaced. In a later chapter this will be dealt with more in detail, but when explosions take place in a chamber which remains intact after the explosion, the situation is very close to the theoretical idealization. Under such circumstances it is usually only possible to make relevant measurements of stress waves travelling in one direction, and one loses the possibility of detecting systematic errors in the data. Great care must, therefore, be taken in scrutinizing the data, but even so the results will usually be slightly inferior in accuracy to those cases in which such systematic errors may be extracted. In the following, a chamber detonation experiment will be discussed in detail.

In experiment F24 charges of varying magnitude were detonated in a 20-m^3 chamber, and the maximum amplitudes of the stress wave created by the explosion in the rock were recorded at different distances from the chamber. The distance from the center of the chamber to the wall normal to the line through the pickups was 2.0 m which then may be taken as the radius a in the theory. This radius is thus the same for all charges. Three charges of each 50 kg TNT and 170 kg TNT and two charges of 500 kg TNT were detonated. Unfortunately the recorded data from the 50-kg charges were so uncertain that only two data remained. These two do not form any justified basis for drawing conclusions, but they fit so well into the overall picture that they have been shown anyway.

Fig. 73 shows the result of the data evaluation and the determined values for λ for the two cases are given in Table VII. The value of λ, which may be obtained from the two data from the 50-kg charge, is also given together with the computed values for τ_0.

It is observed that the general trend of decreasing values of $|\lambda|$ for increasing values of τ_0 is exhibited by the data in complete agreement with the theoretical results.

If now the distance over which the measurements are made (see Fig. 73) are compared with the value of a, one realizes that the values of ξ for which the λ-values are obtained differ from the range $7 \leqslant \xi \leqslant 20$ for which the approximate formulae 9.1 and 9.3, 9.4 and 9.5 are valid. If one, however, disregards this fact for a moment and uses the diagram in Fig. 69 to determine the κ-value, one will obtain a result which is shown in Fig. 74. It is seen that the different charges all give

TABLE VII

Results of F24

Charge (kg)	τ_0	λ	κ	A_0^* (μ-str)	B	A_0 (μ-str)	p_0 (atm)
50	1.4	−1.95103	0.026	880	1.78456	493	38.4
170	3.0	−1.80170	0.026	2136	1.30902	1632	127
500	4.2	−1.69587	0.022	5003	1.29172	3873	301

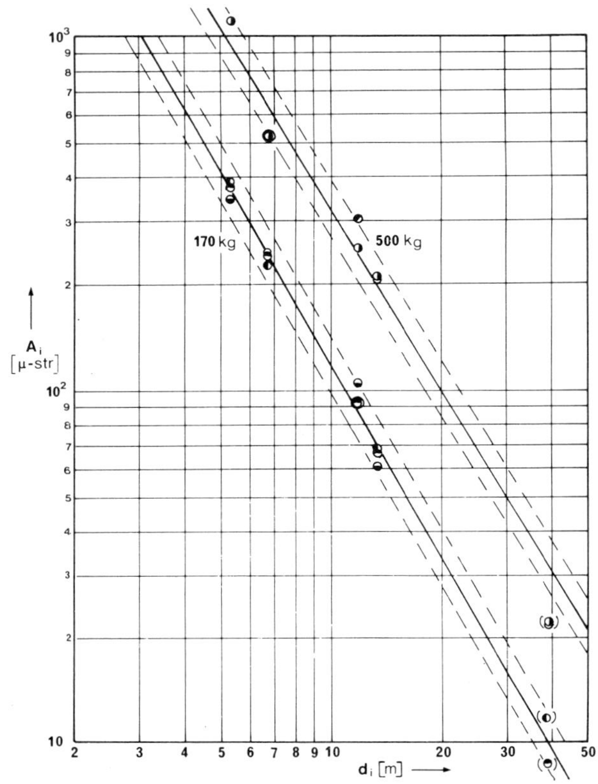

Fig. 73. Result of data evaluation for F24.

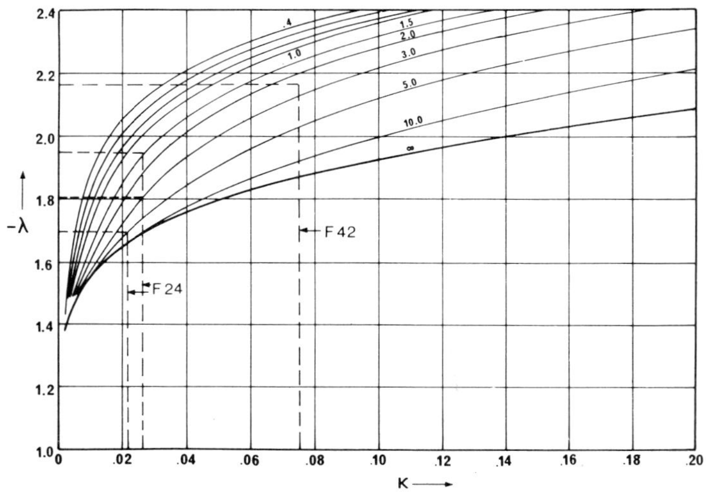

Fig. 74. Determination of κ-values for F24 and F42.

approximately the same κ-value. With $\kappa = 0.026$, the value of B in each case is now computed from the approximate formulae indicated above, and the value of A_0^* at a distance $a = 2.0$ m (computed from the straight lines in Fig. 73) may be reduced to the "true" value of A_0 as shown in Tables VI and VII.

Now attention is drawn to the fact that a conversion factor k(atm/μ-strain) exists for the pickups (determined from calibration of the pickups). For the pickups used in the present experiment $k = 0.07778$ atm/μ-strain. Thus A_0 may be converted into p_0 which then should represent the peak pressure in the chamber during the explosion. This pressure was measured directly by a separate arrange-

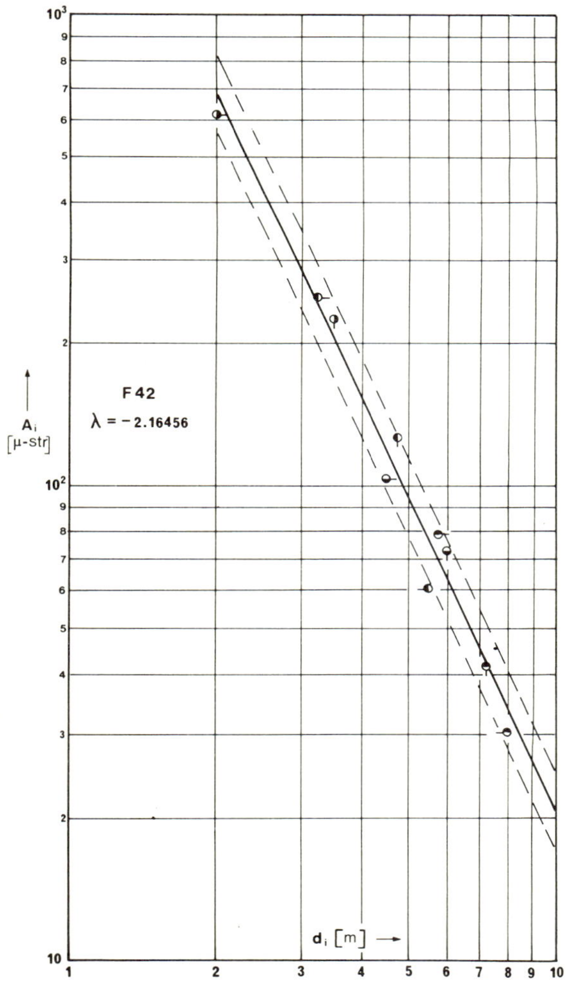

Fig. 75. Result of data evaluation for F42.

ment, and in spite of the uncertainties in these data, the agreement is astonishingly good. (Further comments to the evaluation of the experiment are found in the next chapter.)

It is quite clear that the results from this experiment should be compared with results obtained from a standard row experiment in the same rock massive. F42 is such an experiment where a slightly modified version of the layout in Fig. 58 was used. The results are shown in Fig. 75 and the value for τ_0 was 2.45. Thus this experiment gives according to Fig. 74 a value $\kappa = 0.075$ which is much higher than the value $\kappa = 0.026$ obtained from F24.

This result is very important. The fact that the κ-value obtained from explosions in a chamber is much less than the κ-value obtained from completely confined explosions is a general observation. In both cases the κ-value is obtained from measurements of the propagation of stress waves in the same material. Thus, if κ were a true parameter of the material only, the κ-values should be the same. If on the other hand the contention of J. Kestin (1968) is correct, the κ-value is a parameter of the process, and because the process is very much different at the origin of the stress wave in the two cases, the κ-values should differ as indeed they do. In the chamber explosions the rock material remains more or less intact everywhere, whereas in a confined explosion the rock material is crushed and pulverized close to the detonating charge. With this difference in mind one will be able to argue that the κ-value at a confined explosion should be higher than the corresponding κ-value for the chamber explosion. Thus, the present experiments give strong support to the concept of κ being a parameter of the process and it characterizes the whole system, i.e. the material (rock) as well as the process at the origin of the stress waves. This should be kept in mind in the application of the theory to practical problems.

Chapter 10

EVALUATION OF DATA

INTRODUCTION

The exposition of the experimental investigations of stress waves in rocks given in Chapter 8 shows that the data usually are:

(1) The maximum amplitude of the stress wave (measured in μ-strain).

(2) The duration, t_0, of the positive phase of the stress wave (measured in μ-sec).

(3) The maximum particle acceleration (measured by accelerometers in g's)

(4) The arrival times of the stress waves.

(5) The positions of the pickups (measured in meters).

These data are quantitative in the sense that they are given in terms of numbers which will be inflicted with inescapable errors. Very often qualitative "data" (for instance in the form of geological information) may be drawn upon to extract erroneous data from the whole set of data. Sometimes the data obtained may lead to erroneous conclusions because the total background for the data has not been taken into account. It is the purpose of this chapter to exhibit such cases based on results from actual experiments. It has been presumed that the data evaluation shown in Chapter 9 is well understood.

THE RADIUS *a* OF THE SPHERICAL CAVITY

It has been pointed out previously, that the radius (a) of the spherical cavity with which the explosion is theoretically replaced, is in reality an "imaginary" quantity, and as such cannot be measured. In the previous chapter some considerations were given by means of which this quantity was estimated. Because a is such a crucial quantity in the evaluation of the data, a closer examination of its physical significance is appropriate.

Fig. 76a shows a sketch of the charge (shaded) as it is positioned in the hole before detonation, the sketch being relevant to the 0.5-kg charges usually used in a standard experiment. After detonation a situation as sketched in Fig. 76b will occur. The volume, V, of the enlarged cavity can rather easily be measured, and because its shape can be approximated by a sphere, its radius can be found. If now the same hole is used for another 0.5-kg charge, the situation (and the process taking place during the explosion) will not be the same as the first time. But if the

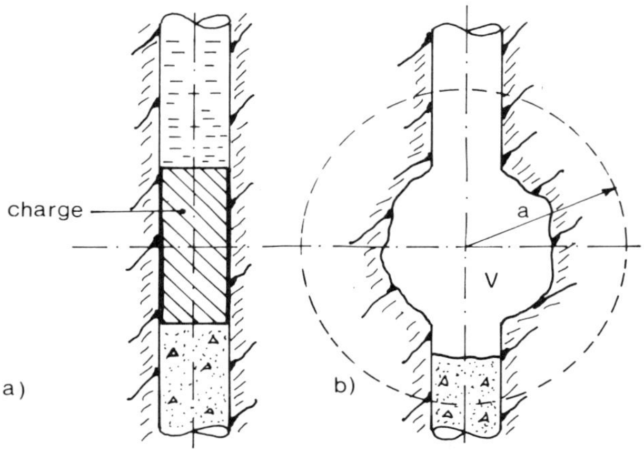

Fig. 76. Charge positioned in the hole before detonation (a) and the hole after the explosion (b).

charge is properly increased, the situation will again be comparable to the first explosion of 0.5 kg. One might now conceive of an experiment where increasing charges were detonated in the same hole, each time measuring the volume, V, of the cavity left by the explosion. The result of such an experiment would depend on several factors, the main ones being: (1) the characteristics of the explosive; and (2) the characteristics of the rock, i.e. its ability to withstand pulverization by the explosive.

It is thus clear, that the problem is rather complicated, but to obtain comparable results, the experiments referred to in this text have almost exclusively been made with one explosive only, which has a rather high rate of combustion (combustion velocity 8000 m/s). An experiment of the above type has been performed in granite, with the result:

$$V = 3.5W \qquad [10.1]$$

where the volume, V, is measured in dm^3, and the weight, W, is measured in kilograms. The factor 3.5 appearing in this equation depends of course both on the rock and the explosive characteristics. It is, however, observed that the radius, r_0, of volume V is proportional to $W^{1/3}$. It seems reasonable to assume that this result may be given a general validity, i.e. the proportionality will occur in most types of rocks.

The next step is to assume that in these cases one will have geometric similarity in the sense that there exists a constant relationship between r_0 and a:

$$a = k \cdot r_0 \qquad [10.2]$$

where the constant k does not change as long as one is using the same explosive in the same rock. This gives rise to the conclusion:

$$a \sim W^{1/3} \qquad [10.3]$$

This result is very important, because it involves the so-called "scaling problem", which will be discussed in detail at a later stage. It is, therefore, imperative to be aware of the uncertainties involved in the deduction of 10.3.

A physical interpretation of radius a may be obtained from the following line of reasoning. If the signals obtained from the pickups close to the charge are examined, it is observed that the initial slope of the signals decreases with increasing distance. The situation is the same as the one shown in Fig. 77 where it is seen that the slope (indicated by dashed lines) changes drastically from $\xi = 2$ to $\xi = 3$ for the case under consideration. Because the input pulse is to be replaced by a rectangular one, it means that the slope at $\xi = 1$ is infinite. If this then is carried over to the experimentally obtained signals, it means that a is the distance at which an imaginary rectangular pulse would be acting so as to create the right slopes at the pickups. This distance can be estimated and for the standard experiments referred to in this text, the value for the 0.5-kg charge detonated the first time would be 0.2 m $\leqslant a \leqslant$ 0.4 m. Thus $a > r_0$ in these cases. However, for the chamber explosions discussed in the previous chapter the situation is different. Here the pressure created by the explosion on the chamber walls is being gradually and very rapidly built up giving already at that point an input pulse with a slope which is not infinite. Therefore, in this case the radius a must be smaller than the dimensions of the chamber. And this is the reason why the apparent discrepancy between the range of values of ξ for the experiment and the corresponding range for the approximate formulae used to determine κ was not a serious one. It should be mentioned that considerations of this type were made in all numerical evaluations exhibited so far in this text, although at the time it was not specifically mentioned. This was done deliberately in order not to unnecessarily complicate the subject.

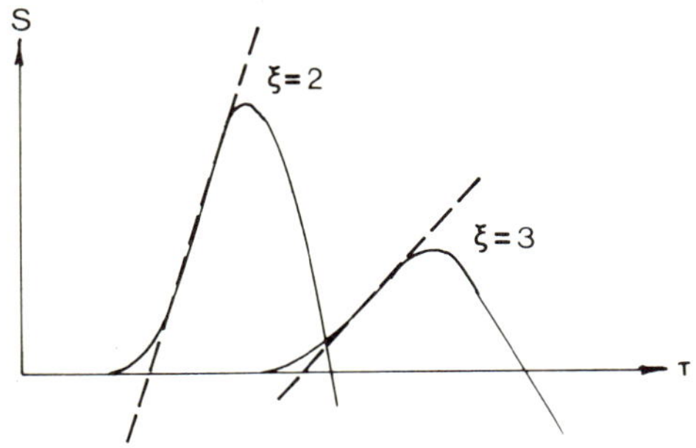

Fig. 77. Theoretical "signals" at $\xi = 2$ and $\xi = 3$ computed with $\kappa = 0.25$ and $\tau_0 = 0.6$.

THE NON-DIMENSIONAL IMPULSE τ_0 OF THE INPUT PULSE

The second parameter in the theoretical approach which is of great importance in the numerical evaluation of the data is τ_0. As exhibited earlier, this parameter is a non-dimensional measure of the impulse of the input pulse. It is, however, also the non-dimensional duration time of the rectangular input pulse which is substituted for the real pulse. In Chapter 9, section *Comparison between theory and experiment,* the determination of τ_0 was discussed in detail. It should, however, be kept in mind that these measurements are the most uncertain ones in experimentations in rocks because they are most sensitive to local inhomogeneities in the rock. Also, imperfections in the way in which the pickups are fixed in the rock will influence the measurement of duration times of the positive phase of the pulse much more than they affect the measured maximum amplitude. Under such circumstances the measurements of the duration time, t_0, of the positive phase of the pulse should always be considered in connection with other information available.

One such piece of information may be obtained from considerations of the combustion of the explosive. Any explosive may be characterized by its combustion velocity v_c. Therefore it may be concluded that the time (Δt) necessary to complete the combustion of a charge will be:

$$\Delta t = \frac{d}{v_c} \qquad [10.4]$$

where d is a characteristic length of the charge which will be proportional to $W^{1/3}$, W being the weight of the charge. Thus, comparing Δt for two different charges, W_1 and W_2 one will find:

$$\frac{\Delta t_1}{\Delta t_2} = \left(\frac{W_1}{W_2}\right)^{1/3} \qquad [10.5]$$

If the explosion takes place in exactly the same way in the two cases, one may further conclude that the duration times, $t_{0.1}$ and $t_{0.2}$, of the input pulses in the two cases will be related in the same way, thus:

$$\frac{t_{0.1}}{t_{0.2}} = \left(\frac{W_1}{W_2}\right)^{1/3} \qquad [10.6]$$

This result is now used in connection with 10.3 and 9.9, and one arrives at the simple conclusion: *for two different charges W_1 and W_2 the values of τ_0 will be the same if the conditions are ideal and similarity may be assumed.*

This is a very important result, and it must be confronted with experimental evidence. Any departure from this result may now be interpreted in terms of local variations in the radius, a, and the duration times, τ_0, two quantities which greatly influence each other. An examination of experimental data will show the details.

RELATION BETWEEN THE CHARGE FACTORS AND THE RADIUS a

It will be recalled that a "charge factor" β was introduced in Chapter 8 to eliminate the systematic deviations between data from charges of equal magnitude (W) but in different positions. One observes that the effect of such a charge factor in the diagram for the attenuation curve (doubly logarithmic scale) will be to move all data from one detonation an equal distance in the vertical direction as indicated in Fig. 78. The argument for this procedure rested on the observation that two equal charges in two different holes do not necessarily create the same peak value of the input pulse. The later considerations have drawn the volume, V, created by the explosion into the picture. This volume depends on the characteristics of the rock which even in the same rock massive may differ from one charge hole to the other. There is thus a relation between the peak pressure created by the explosion and the value of the radius, a. This means that the charge factors may be replaced by a "distance factor" whereby the original data in Fig. 78 (○) instead of being displaced vertically to their positions (◐) will be displaced horizontally to the positions (●). (In both cases the translated data will group themselves around the commonly determined straight line according to the best-fit procedure.)

The "distance factor" represents, however, nothing but deviations in the radius a.

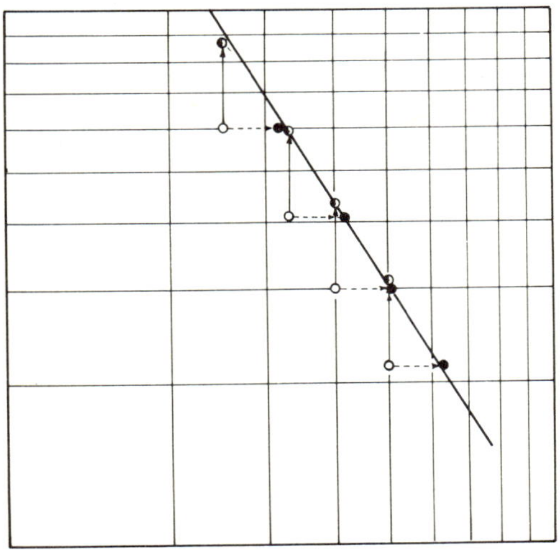

Fig. 78. Effect of "charge factors" and "distance factors". ○ = original data, ◐ = vertically displaced data (charge factor), ● = horizontally displaced data (distance factor).

Thus, the value of a to be used for each charge to make the distance dimensionless is obtained from geometrical considerations (Fig. 78):

$$a = a_0(\beta)^{-1/\lambda} \tag{10.7}$$

where: β is the charge factor determined from the previously exhibited method of evaluation (Chapter 8), λ is the slope of the attenuation curve determined by the same procedure, and a_0 is a "mean" value of a to be adopted for the experiment. (Comments on this value will be given in connection with worked out examples.)

If this is used in the evaulation of data from an experiment, it means that data from charges of different magnitudes may be evaluated together assuming that the conclusion following 10.6 holds. Details of such an evaluation will now be given for an actual experiment.

EVALUATION OF F47 (NUMERICAL EXAMPLE)

The experiment F47 is a standard row experiment using the layout exhibited in Fig. 58. The rock was granite with a signal velocity of 6000 m/s and first a charge of 0.5 kg PV was detonated in each of the holes, A, B, ... H. Then the entire experiment was repeated with 2.5 kg PV detonated in the same holes. The data obtained from the pickups gave the maximum amplitudes A_i(μ-strain) measured at the distances, d_i(m), as shown in Table IX. These data will be referred to as the "raw data", and are plotted in a doubly logarithmic scale in Fig. 79 for the 0.5-kg charges. Because it is important to keep track of each detonation as well as each pickup, the legend in Table VIII is used.

Already at this stage certain characteristics of the data are obvious. In Fig. 79 it is observed that all data obtained from charges detonated in holes E, F, G and H lie below the corresponding data obtained from detonating charges in holes B, C and D. In addition, it is observed that the measured values of t_0 (the duration time of the positive phase of the signal measured in μ-strain) differ very much for data obtained when the wave is moving in different directions. Such tendencies in the data may indicate a major surface of discontinuity in the rock between hole E and

TABLE VIII

Legend

Detonation in	A	B	C	D	E	F	G	H
Legend	◓	◒	◐	◑	◕	◔	◕	◔
Pickup	1	2	3	4	5	6		
Legend	⫯	o–	⫰	–o	⌽	–o–		

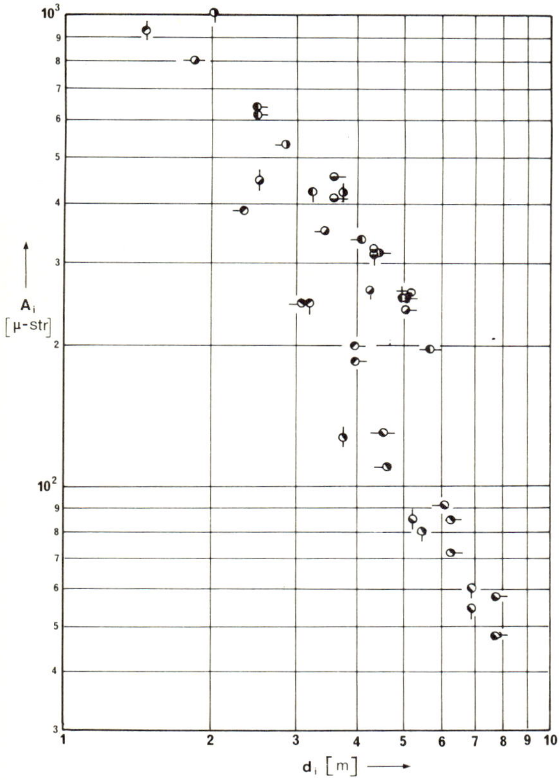

Fig. 79. Raw data from F47 for 0.5-kg charges.

pickup 6 (see Fig. 58). Even though this would indicate that data from the shots in E, ... H, should be disregarded, the evaluation will be carried out for all relevant data, with the hope that the evaluation will show this discrepancy in the end.

The first step in the evaluation is to decide which data are relevant. Remembering that the approximate formulae [9.1] which will be used to determine κ (once λ is determined) is based on the assumption that $10 \leqslant \xi \leqslant 20$ with the possible extrapolation to both smaller and larger values of ξ, it is clear that some idea of the radius, a, must be found. Based on earlier evaluation it is therefore assumed here that a "mean" value will be $a_0 = 0.3$ m for a charge of 0.5 kg. (This choice is not crucial at the moment, but it will be later.) If it is assumed that only data for which $6.7 \leqslant \xi \leqslant 27$ will be used to determine λ, then only data for which the distances, d_i, satisfy the following condition are permissible:

$$2.01 \text{ m} \leqslant d_i \leqslant 8.1 \text{ m} \qquad [10.8]$$

TABLE IX

All A_i data from experiment F47

Charge in	d_i	A_i		t_0	β	a	τ_0	Remarks
B	3.58	410	455	160				
◒	4.34	310	320		0.94077	0.3131	1.5	
(0.5 kg)	5.18	259						
C	2.49	615	640	100				
◐	3.25	421						
	4.09	335			1.04939	0.2901	1.0	
	4.98	250						
(0.5 kg)	5.67	195						
D	1.26	(1580)	(1800)	140				
◑	2.02	1030						
	2.86	530			1.01293	0.2973	1.4	
	3.75	420						
(0.5 kg)	4.44	320	210					
E	3.96	184	199					
◔	3.20	244						
	2.36	388			1.85878	0.1945	2.3	Must be excluded
	1.47	(930)						
(0.5 kg)	0.78		(1590)	150				
F	5.02	238	255					
◕	4.26	262						
	3.42	350			1.16735	0.2693	4.6	Must be excluded
	2.53	448		410				
(0.5 kg)	1.84		(800)					
G	6.24	72	85					
◔	5.48	80						
	4.64	109			2.45042	0.1604	3.7	Must be excluded
	3.75	127						
(0.5 kg)	3.06	244	214	200				
H	7.72	48	58					
◕	6.96	60	55					
	6.12	91			2.46916	0.1595	3.8	Must be excluded
	5.23	85						
(0.5 kg)	4.54	130	121	200				
C	2.49	(1160)	(1250)					
	3.25	(810)						
	4.09	610		180	0.59793	0.4297	1.3	
	4.98	486						
(2.5 kg)	5.67		272					
F	5.02	491	479					
	4.26	570		570	0.52897	0.4687	3.6	Must be excluded
	3.42	781						
(2.5 kg)	2.53	(1250)						
G	6.24	117	122					
	5.48	120						
	4.64	161		300	1.64608	0.2118	4.3	Must be excluded
	3.75	202						
(2.5 kg)	3.06	(332)	(310)					

If furthermore the validity of 10.3 is assumed, then for 2.5-kg charges $a_0 = 0.51$ m must be used and the corresponding conditions for these charges will be:

$$3.43 \text{ m} \leqslant d_i \leqslant 13.85 \text{ m} \qquad [10.9]$$

In the table displaying the "raw data", those which are thus eliminated in the evaluation have been put in parentheses. The rest of the data are then evaluated through the procedure exhibited in Chapter 8, with the result that:

$$\lambda = -1.43091 \qquad [10.10]$$

Values for the charge factors, β, have been obtained by using the mean value of the shots in B, C and D only. They are given in Table IX as are the corresponding values of a computed from 10.7. Then using 9.9 one may compute the values of τ_0 for each charge. These values are also given in Table IX.

Now the results will have to be examined. The charge factors express what was previously obvious: the data from charges in E, F, G, H are too low. The values of τ_0 show, however, that these charges give values which far exceed those from the B, C, D side, which all show $1.0 < \tau_0 < 1.5$. Thus the evaluation shows that it has been based on the wrong set of data, because only data with reasonably similar values of τ_0 can be evaluated together.

The data are now re-evaluated taking only data from the shots in B, C and D into account. The result is shown in Table X. It is seen that the values of β and a

TABLE X

Selected A_i data from experiment F47

Charge in	d_i	β	a	ξ	A_i		τ_0
B	3.58			11.38	410	455	
◒	4.34	0.92932	0.3147	13.79	310	320	1.53
(0.5 kg)	5.18			16.46	259		
C	2.49			8.57	615	640	
◐	3.25			11.19	421		
	4.09	1.05136	0.2904	14.08	335		1.03
	4.98			17.15	250		
(0.5 kg)	5.67			19.52	195		
D	1.26			4.26	(1580)	(1800)	
◑	2.02			6.84	1030		
	2.86	1.02349	0.2955	9.68	530		1.42
	3.75			12.69	420		
(0.5 kg)	4.44			15.03	320	210	
C	2.49			5.83	(1160)	(1250)	
⬖	3.25			7.61	(810)		
	4.09	0.58098	0.4272	9.57	610		1.26
	4.98			11.66	486		
(2.5 kg)	5.67			13.27		272	

obtained now differ only slightly from the previous values. The values of τ_0 are close enough, as will be shown later. The values of a make it possible to compute the ξ-values for each data and these satisfy the previous conditions very well.

Fig. 80 shows the result of the evaluation. (Observe that in the figure ◐ has been replaced by ◆ to distinguish between data in C from 0.5-kg and 2.5-kg charges). The value of λ for the attenuation curve is now:

$$\lambda = -1.53619 \qquad 1.03 \leqslant \tau_0 \leqslant 1.53$$
$$\text{mean value: } \tau_0 = 1.31 \qquad [10.11]$$

The λ-value is rather different from the value obtained in 10.10. In Fig. 80 also the data points which were omitted from the numerical evaluation have been plotted (in parentheses) and they are seen to fit very well. Attention is especially drawn to the values closest to $\xi = 1$ which should always lie below the straight line for reasons which were explained in dealing with the theoretical solution.

It should be mentioned that any determination of κ from experimentally evaluated λ-values should be confirmed by repeated experiments in the same rock mas-

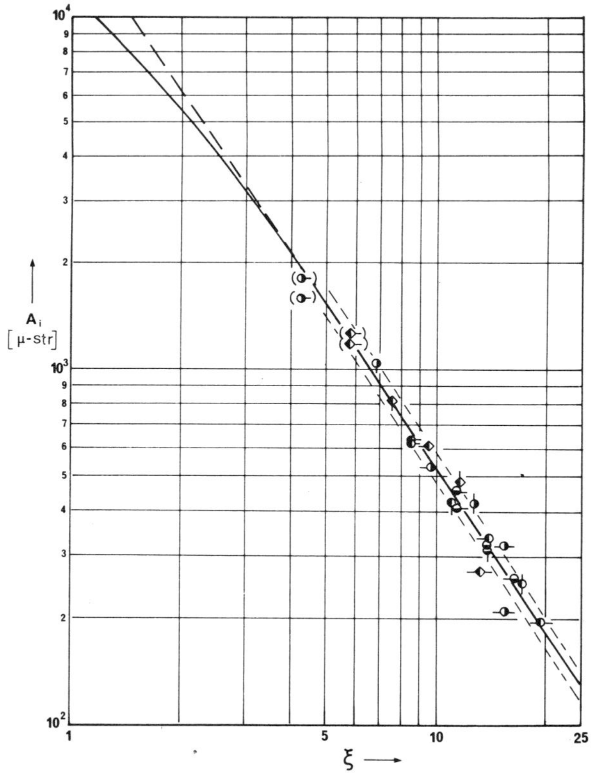

Fig. 80. Result of the evaluation of F47.

sive. In this case an identical experiment, F50, was made on a different location in the same rock. The result is here only quoted without details:

$$\lambda = -1.51852 \qquad 1.39 < \tau_0 < 1.97, \qquad \text{mean value: } \tau_0 = 1.63 \qquad [10.12]$$

The κ-value for the rock is now determined from the diagram in Fig. 69, an enlarged portion of which is shown in Fig. 81. It is seen that variation in τ_0 for each experiment gives an interval marked with bold lines within which the κ-value will lie. Using the mean value of τ_0 for each experiment, they both determine κ = 0.006, the agreement being remarkably good.

Now one may turn to the value of A_0^* created by the 0.5-kg charge at a = 0.3 m. For the two experiments one will have the values given in Table XI. From the approximate formulae (9.3, 9.4 and 9.5), the reduction factor B may be computed whereupon the real values of A_0 at a = 0.3 m will be as given in the table with its corresponding peak input pressure, p_0. These deviate 14.8% from their mean value.

Reconsidering the procedure followed here it will be observed that as far as the determination of the λ-values are concerned the chosen "mean value" for a is not crucial. When it comes to the value of τ_0 to be used it is! It is therefore very important to have more than one experiment to rely on when κ is to be determined. In order to bring out to what extent the value of a will influence the results it should be mentioned that if in the experiments above the mean value of a had been chosen as

a = 0.33 m for F47 instead of a = 0.3 m

a = 0.29 m for F50 instead of a = 0.3 m

the values of A_0 (or p_0) would have been identical for the two experiments, and

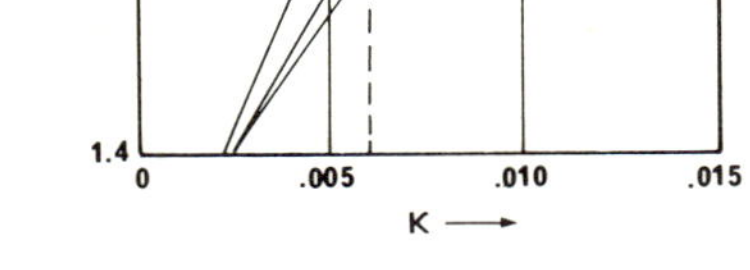

Fig. 81. Diagram for determination of κ from F47 and F50.

TABLE XI

Result of F47 and F50

Exp.	A_0^*	τ_*	B_∞	B	A_0	p_0(atm)
F47	18158	0.21302	1.30053	1.46865	12364	961
F50	12468			1.35968	9170	713

TABLE XII

A_i data from experiment F46

Charge in	W(kg)	A_i		d_i	λ	ξ_i	β	ξ_i(corr.)
F	0.5	(7)	(9)	14.95		37.4		
	◕	(14)		11.71	−1.8542	29.3		
		17		9.81		24.5		
		29		7.85	$m = \pm 9.8\%$	19.6		
		32		6.90		17.3		
	1.5	16	16	14.95		25.9		
	◪	27		11.71	−1.9353	20.3		
		36		9.81		17.0		
		55		7.85	$m = \pm 2.1\%$	13.6		
		73		6.90		12.0		
	2.5	22	19	14.95		21.9		
	▲	34		11.71	−1.7770	17.1		
		46		9.81		14.3		
		67		7.85	$m = \pm 5.7\%$	11.5		
		79		6.90		10.1		
	5.0	32	30	14.95		17.3		
	◆	57		11.71	−1.7582	13.6		
		70		9.81		11.4		
		105		7.85	$m = \pm 7.6\%$	9.1		
		117		6.90		8.0		
E	0.5	14	13	10.98	−2.2855	27.5		26.9
	◕	30		7.74		19.4	0.96314	18.9
	1.5	27	26	10.98		19.0		18.6
	◩	55		7.74		13.4		13.1
		76		5.84	−2.1304	10.1	0.95929	9.9
		(290)		3.88		6.7		6.6
		(405)		2.93	$m = \pm 15.0\%$	5.1		5.0
	2.5	35	31	10.98		16.1		16.8
	▲	62		7.74	−1.9734	11.3		11.9
		85		5.84		8.5	1.08233	8.9
		(300)		3.88		5.7		5.9
		(410)		2.93	$m = \pm 15.0\%$	4.3		4.5

the values of κ would still have been the same. The conclusion to be drawn from this is that still more information on radius a is needed (see Chapter 11).

EVALUATION OF F46 (NUMERICAL EXAMPLE)

The experiment to be presented in detail here was performed in the same rock massive as F47 and F50. The rock is granite and has a signal velocity c_s = 6000 m/s. The layout was similar to the one shown in Fig. 58 with the modification that more pickups were introduced in the middle section and the distances were increased because of the intended use of increasing charges in each hole. The charge weights were W = 0.5, 1.5, 2.5, 5, and 10.25 kg and the intention was to detonate such charges in the order of increasing W in each hole. The experiment took place in early spring. There was frost in the ground which on one hand introduced practical difficulties in performing the experiment and on the other (due to ice in all cracks and possible discontinuous surfaces in the rock) changed the characteristics of the medium transmitting the stress waves. This led to loss of data due to severe weather conditions and only results from charges in E and F were obtained in sufficient

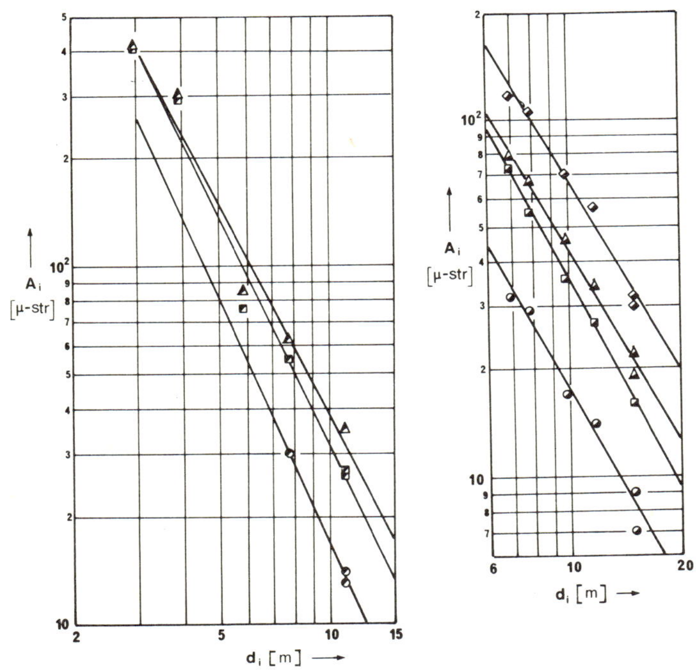

Fig. 82. Results of the first attempt to evaluate the data from F46.

numbers to permit evaluation. These data are given in Table XII as maximum amplitudes, A_i(μ-strain), measured at distances, d_i(m). The table also gives the legend used in the following diagrams to distinguish between shots of different charge magnitudes.

If now all these data are indiscriminately evaluated for each charge by determining the straight lines which fit through them, the values for the slopes, λ, of these lines will be as given in Table XII. The standard deviations, m, is given in percent and the result is shown in Fig. 82. Apparently the results are not bad, the general tendency of decreasing values of $|\lambda|$ with increasing values of W being present in accordance with theoretical predictions. It is, however, found upon closer examination of the data, that an apparent discontinuity is present in the data from shots in E. This causes the λ-values from these shots to differ considerably from those obtained from shots in F. The general conclusion must therefore be that λ has not been adequately determined.

The next step in improving the evaluation would then be to evaluate the data for equal shots (same W) in E and F using the previously introduced procedure. The result of this procedure is shown in Fig. 83 which clearly exhibits trends in the results which are not only contrary to the theoretical predictions but also to physical reality.

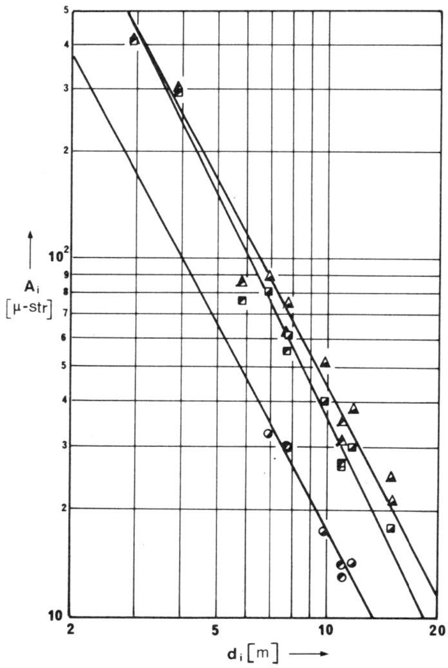

Fig. 83. Result of the second attempt at evaluating the data from F46.

If now the result in 10.3 is considered, the distances d_i should be made dimensionless with the radius, a, which is proportional to $W^{1/3}$. Thus if a is equal to 0.4 m for $W = 0.5$ kg, then the value of a to be used for the other charges would be:

$$\left.\begin{aligned}
&\text{for } W = 0.5 \text{ kg:} \qquad a = \left(\frac{0.5}{0.5}\right)^{1/3} \cdot 0.40 = 0.40 \text{ m}\\
&\text{for } W = 1.5 \text{ kg:} \qquad a = \left(\frac{1.5}{0.5}\right)^{1/3} \cdot 0.40 = 0.577 \text{ m}\\
&\text{for } W = 2.5 \text{ kg:} \qquad a = \left(\frac{2.5}{0.5}\right)^{1/3} \cdot 0.40 = 0.684 \text{ m}\\
&\text{for } W = 5.0 \text{ kg:} \qquad a = \left(\frac{5.0}{0.5}\right)^{1/3} \cdot 0.40 = 0.862 \text{ m}
\end{aligned}\right| \quad [10.13]$$

The non-dimensional distances, ξ_i, for each data can be computed as:

$$\xi_i = d_i/a \qquad [10.14]$$

where for the particular shot in question the value of a to be used is taken from 10.13. These values of ξ_i are also to be found in Table XII. In Fig. 84 the data from all shots in F is now plotted with ξ_i as the absissa, taking only account of data for which:

$$7 \leqslant \xi_i \leqslant 27 \qquad [10.15]$$

This means that data in the table are present which should not be taken into account

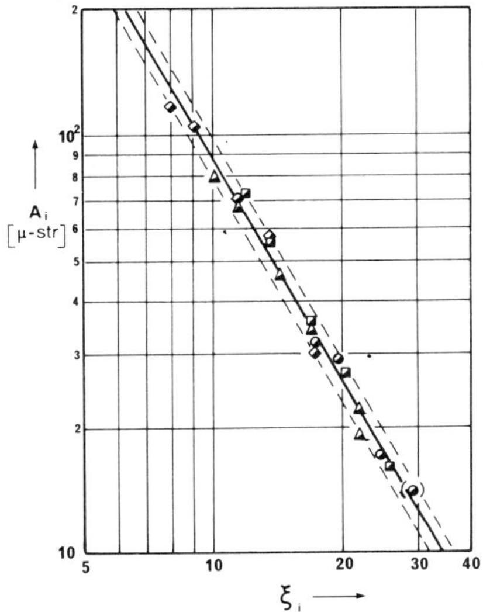

Fig. 84. Results of the evaluation of relevant data from F46. Charges in F.

when λ is determined. These are put in parentheses in the tables, but it is stressed that in the earlier evaluations no attention was paid to this possibility and thus *all* data were indiscriminately used.

It appears that all data gather around a straight line which can be found by a best-fit method. This gives:

$$\lambda = -1.77754 \qquad m = \pm 7.5\% \qquad [10.16]$$

where m is the standard deviation. It is seen that almost all data gather within the band of ± 10% around the line. The results thus represent a typical case where relation 10.3 is satisfied with an astonishing degree of accuracy. The obtained value of λ is based on all relevant data from shots in F.

The same procedure is tried for the data obtained from shots in E. They do not gather as nicely as for shots in F. They are therefore evaluated using the ordinary procedure with introduction of charge factors. These are given in Table XII, and they are used to modify the values of a used to determine the ξ_i's. One thus obtains:

$$\left.\begin{array}{ll} \text{for 0.5 kg:} & a = 0.4089 \text{ m instead of originially: } a = 0.400 \text{ m} \\ \text{for 1.5 kg:} & a = 0.5912 \text{ m instead of originally: } a = 0.577 \text{ m} \\ \text{for 2.5 kg:} & a = 0.6530 \text{ m instead of originally: } a = 0.684 \text{ m} \end{array}\right| \quad [10.17]$$

It is seen that only minor changes in a are made. With these values the corrected values of ξ_i are computed and displayed in Table XII. The result of the procedure is shown in Fig. 85 and the value of λ obtained is:

$$\lambda = -1.70556 \qquad [10.18]$$

which at least is close to the value (10.16) from the evaluation of the data obtained with charges in F.

In Figs. 84 and 85 also the data which were not considered have been plotted (in parentheses). For shots in F they fit very well, for shots in E they do not as is easily recognized from Fig. 85. It seems as if a discontinuity surface has existed between pickups 6 and 7, but this cannot be recognized from the data produced by shots in F. Furthermore the measured duration times, t_0, differ radically from the values usually obtained with the same charges in the same rock, which can easily be attributed to the presence of ice in the rock. Thus a further evaluation based on these data does not seem worthwhile due to the inherent uncertainties.

However, the results exhibited so far have given the following pieces of information:

(1) The assumed proportionality between a and $W^{1/3}$ has been strongly supported.

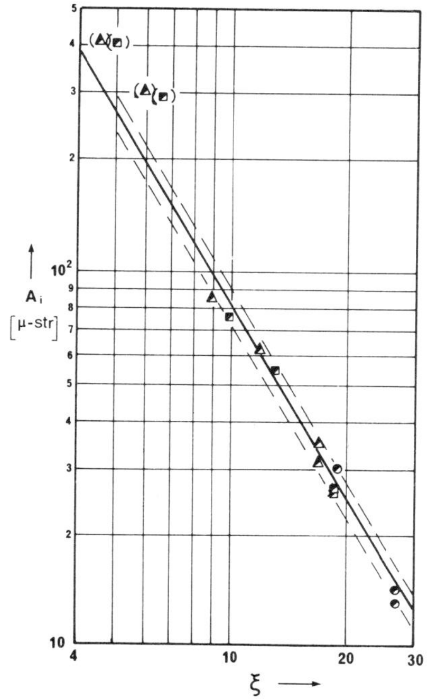

Fig. 85. Result of the evaluation of data from F46. Charges in E.

(2) The results show that even small deviations in a have large effects, but that this may be taken care of by the conversion of "charge factors" β to appropriate values of a according to 10.7.

(3) The values of $|\lambda|$ in F46 are seen to be much larger than the corresponding values from F47 and F50 obtained in the same rock massive. One might therefore be tempted to conclude that a rock shows a greater damping when it is frozen than when it is dry. Attention should, however, be paid to the uncertainty of the results in F46, upon which such a conclusion rests.

Chapter 11

DISPERSION AND PARTICLE ACCELERATION: APPLICATION IN THE EVALUATION OF DATA

INTRODUCTION

So far the numerical evaluation of data has been used to lend strong credence to the applicability of the theoretical results. It has, however, all along been pointed out that the radius, a, of the "imaginary sphere" with which the explosion is replaced remains unknown. Of course, the previous discussions have to some extent given an idea of its order of magnitude, and the relation between the correct values of a could be extracted from the data. However, the "exact" value of a has not been found, and it should also be stressed, that so far only a limited part of the experimentally obtained information has been drawn upon in the considerations. From the signals obtained, only the maximum amplitudes, A_i, of the stress waves have been considered, whereas also a physical reality as the dispersion ought to be drawn into the picture. Now, the dispersion is not easily described, but to some extent the maximum value of the slope of the signal in its first rising phase, $(\partial S/\partial\tau)_{max}$, may be conceived of as a measure of the dispersion. In addition one may have cases where the sensors used in the pickups measure particle accelerations. Thus, a complete picture of the wave propagation in rock may only be obtained when all these quantities can be interrelated theoretically. It is the scope of this chapter to exhibit how this is brought about.

DISPERSION

Returning for a moment to Chapter 7, section *The arbitrary input pulse,* one finds the attenuation curves for $(\partial S/\partial\tau)_{max}$. Such curves are exhibited in Figs. 53 and 54. It is seen that these curves also exhibit the characteristic feature of being almost straight lines in the range $7 \leqslant \xi \leqslant 20$. It is, therefore, natural to attempt a simplified approach for exhibiting the theoretical results by the introduction of approximate formulae, which within a certain range of accuracy (e.g. 1%) exhibit the essential behavior of the solution. The idea is, in other words, to do exactly the same for the dispersion as was done for the maximum stress amplitude (or the maximum strain amplitude) in Chapter 9. Fig. 86 shows the theoretical curves of $(\partial S/\partial\tau)_{max}$ for $\kappa = 0.05$ with $\tau_0 = \infty$ (solid curve) and $\tau_0 = 0.4$ (dash-dotted curve).

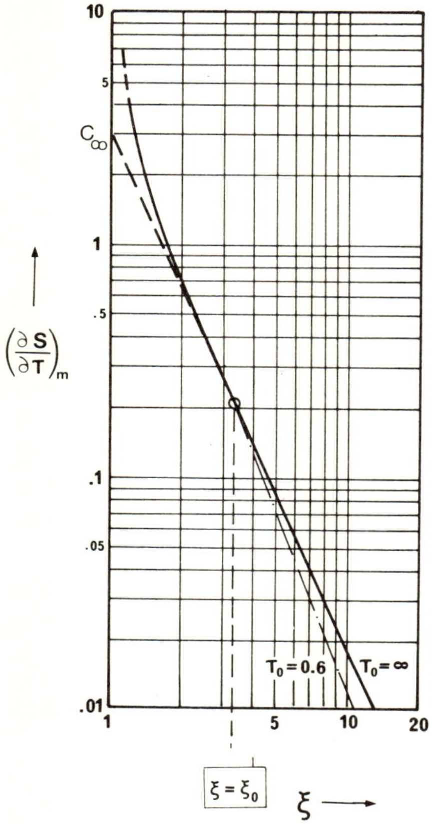

Fig. 86. $(\partial S/\partial \tau)_{max}$ as a function of ξ. Solid curve: $\tau_0 = \infty$, dash-dotted curve: $\tau_0 = 0.4$, dashed curve: approximation for $\tau_0 = \infty$. $\kappa = 0.05$.

If the first one is approximated by a straight line, this line (dashed curve) will give the value C_∞ at $\xi = 1$ where the theoretical curve tends to infinity. If the slope λ_a of the straight-line approximation to the theoretical curve is known, as well as the value C_0 of this line at $\xi = 1$ for any value of κ and τ_0, the theoretical solution in the range $7 < \xi < 20$ is known. Because, however, the straight lines for $\tau_0 \neq \infty$ intersects the line for $\tau_0 = \infty$ at $\xi = \xi_0$, it is better to give ξ_0 as a function of κ and τ_0 instead of C_0. This latter quantity is therefore here given as $C_0 = C_\infty$ for $\tau_0 = \infty$ only.

One will thus have the following set of approximate formulae expressing the theoretical results:

$$\left.\begin{aligned}\lambda_{a,\infty} &= -3.01114\kappa^{0.09568}\\ \lambda_a - \lambda_{a,\infty} &= -0.54337\mathrm{e}^{-(\alpha\tau_0)^2}\\ \alpha &= 0.17109/\sqrt{\kappa}\end{aligned}\right\} \quad [11.1]$$

Here $\lambda_{a,\infty}$ is the slope for $\tau_0 = \infty$, and the value of $C_0 = C_\infty$ for this value of τ_0 will be given by:

$$C_\infty = \frac{0.11162}{\sqrt{\kappa}} + 0.90951\mathrm{e}^{-6\kappa} \quad [11.2]$$

These formulae can be shown in the form of diagrams. Fig. 87 shows λ_a as a function of κ with τ_0 as a parameter. It is noticed that λ_a increases with decreasing values of τ_0 for a given value of κ, and that it also increases with increasing values of κ for a given value of τ_0, features which also are supported by experimental evidence.

Fig. 88 shows C_∞ as a function of κ. It is seen that $C_\infty \to \infty$ at $\kappa = 0$, which raises the question of how these approximate formulae behave when $\kappa \to 0$, i.e. in the case of an elastic medium. It will be remembered from the previous exposition of the wave propagation in an elastic medium, that in the case of an impulsively started input pulse, (i.e. an initially infinite value of $\partial S/\partial\tau$) the value of $(\partial S/\partial\tau)_{max}$ will remain infinite throughout its propagation (i.e. there will everywhere be observed an initial step in the signal). Thus, $\lambda_a = 0$ for all values of τ_0, and $C_\infty = \infty$ for $\kappa = 0$.

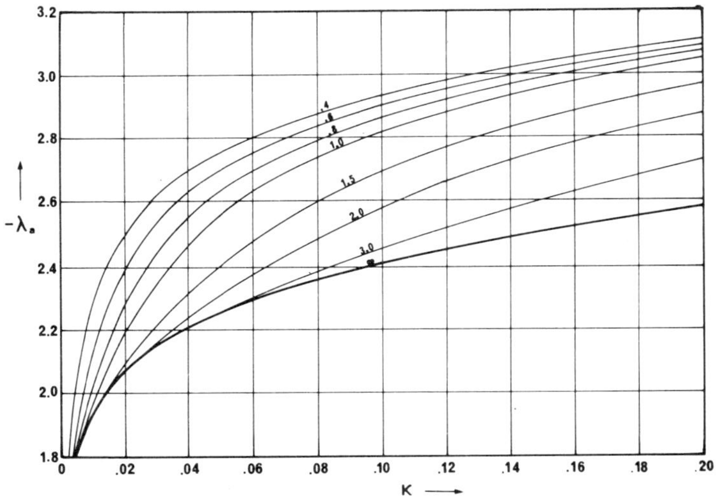

Fig. 87. The slope, λ_a, of the attenuation curves for $(\partial S/\partial\tau)_{max}$ as a function of κ with τ_0 as parameter.

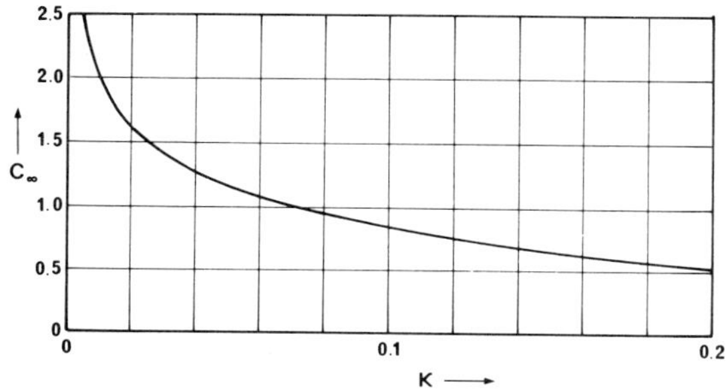

Fig. 88. C_∞ plotted as a function of κ.

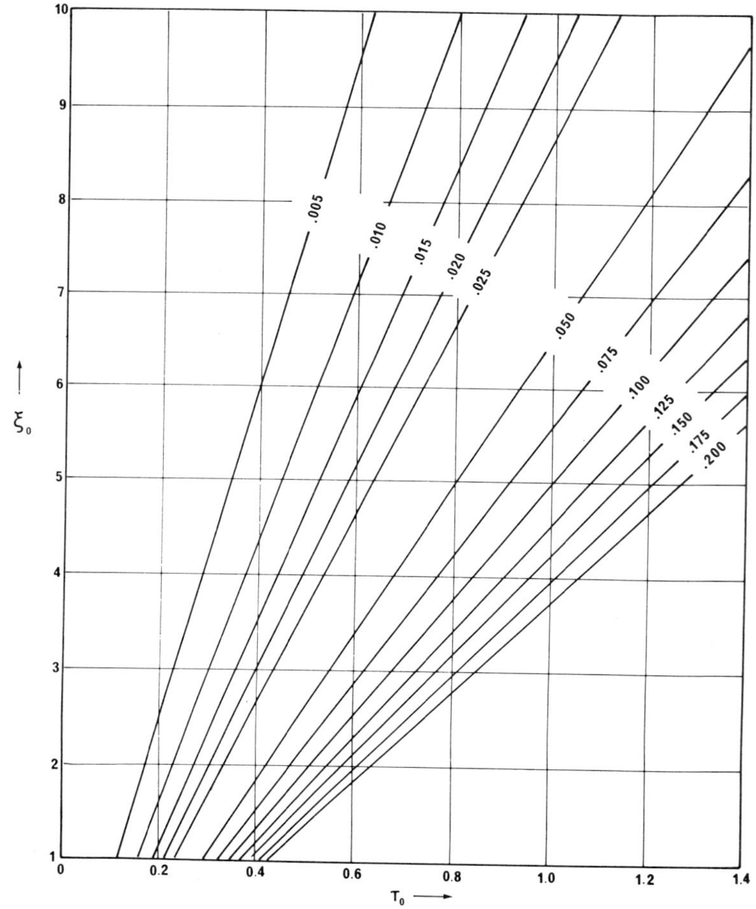

Fig. 89. ξ_0 plotted as a function of τ_0 with κ as a parameter.

This is exactly the behavior of the approximation formulae 11.1 and 11.2. It is clear that the value of ξ_0 under these conditions remains undefined.

The last quantity to be determined is ξ_0, which can be obtained from the following set of formulae:

$$\left.\begin{aligned} \xi_0 &= k\tau_0 + b_\xi \\ k &= 2.7046\kappa^{-0.35619} \\ b_\xi &= -1 - 24\kappa\,[1 + (24\kappa)^2]^{-3/2} \end{aligned}\right| \quad [11.3]$$

Fig. 89 shows a plot of ξ_0 as a function of τ_0 with κ as a parameter. It is observed that for $\kappa = 0$, the value of $k = \infty$ indicating that for an elastic medium the value of ξ_0 remains undetermined which is in complete agreement with the previous statement. It should, however, be observed that the asymptotic formulae 11.1, 11.2 and 11.3 are based on the theoretical solution in the interval $7 \leqslant \xi \leqslant 20$, and that this fact should be born in mind when they are applied.

COMPLETE EVALUATION OF AN EXPERIMENT (F57)

The implications of the theoretical results and their application in the evaluation of experimental data may most easily be explained in connection with the evaluation of an actual experiment. For this purpose the data from experiment F57 is drawn to attention. The layout was that of a standard row experiment as shown in Fig. 58. Due to practical difficulties, two different explosives had to be used, a situation which was thought unfortunate at the time, but which here will permit a deeper insight into the applicability of the theoretical results.

The two explosives were Pentrit wax (PV) and Ammongelit (Am) which exhibit combustion velocities of v_c = 8000 m/s and 2000–6000 m/s, respectively. This means, however, that they also will exhibit different values of τ_0, as will become apparent later. From purely empirical considerations it was decided to use charges of 0.625 kg Am and the standard charges were 0.5 kg PV. With the insight gained so far into the mechanism of wave propagation in rocks, it is thus clear that data from these different types of charges must be evaluated separately.

In the experiment under consideration one important feature was added. Pick-ups 2 and 5 had built into them accelerometers, which measured particle accelerations independently of the strains which were measured by means of strain gages. Thus the data appear as two separate sets for the two different types of charges, a fact which will permit conclusions to be drawn beyond what has hitherto been shown.

Table XIII gives the data obtained from the usual charges of 0.5 kg PV in holes

TABLE XIII

A_i, C_i and D_i data from experiment F57 (0.5 kg PV)

Charge in	d_i(m)	A_i(μ-str)	D_i(m/s^2)	C_i(μ-str/μs)	β	a(m)	ξ_i
C	2.05	490		10.7			5.113
	2.65	245	1540	4.4			6.609
	3.50	193		2.7	0.99608	0.40094	8.729
	4.40	150		2.28			10.974
	4.95	126	400	1.25			12.346
	5.80	85		0.68			14.466
D	.75	(1280)		72.0			1.879
	1.35	(800)	5700	14.2			3.383
	2.20	400		7.05	1.00394	0.39907	5.513
	3.10	255		3.55			7.768
	3.65	190	870	2.46			9.146
	4.50	123		1.30			11.276
F	5.40	54		0.64			14.837
	4.80	98	373	1.19			13.189
	3.95	184		2.56	1.17214	0.36395	10.853
	3.05	235		3.96			8.380
	2.50		1550	4.90			6.869
	1.65	(650)		10.00			4.534

C, D and F. The data consist of the maximum strain amplitude, A_i, measured in μ-strain, the dispersion data, C_i, measured in μ-strain/μ-seconds, the maximum acceleration amplitude, D_i, measured in g's and the distances, d_i, from the center of the charge to the points where these data were obtained.

The first step in the evaluation is now to use the procedure described in Chapter 8, section *Data evaluation for a standard experiment,* whereby the charge factors, β, are obtained as shown in Table XIII. (It should be noted, that during this procedure the data in parentheses in Table XIII are not considered, because they are obtained too close to the charge.) The procedure also gives the slope, λ, of the attenuation curve, whereby the charge factors, β, by means of 10.7 may be converted into the proper value of a for each shot, with a mean value of $a = a_0 = 0.4$ m. As pointed out before, this particular choice of a is not crucial at this stage in the evaluation. It affects which data should be kept out of consideration during the evaluation in view of the formulae, which later will be applied to the results. These formulae are valid for values of the non-dimensional distance, ξ, lying within the range $7 \leqslant \xi \leqslant 20$–$30$. In the present case all data have been deleted for which $\xi < 5$, and the proper value of a to be used will be discussed at a later stage. With the values of a thus obtained and given in Table XIII, the values of ξ_i can readily be determined. The result of this evaluation is shown in Fig. 90, in which it is seen

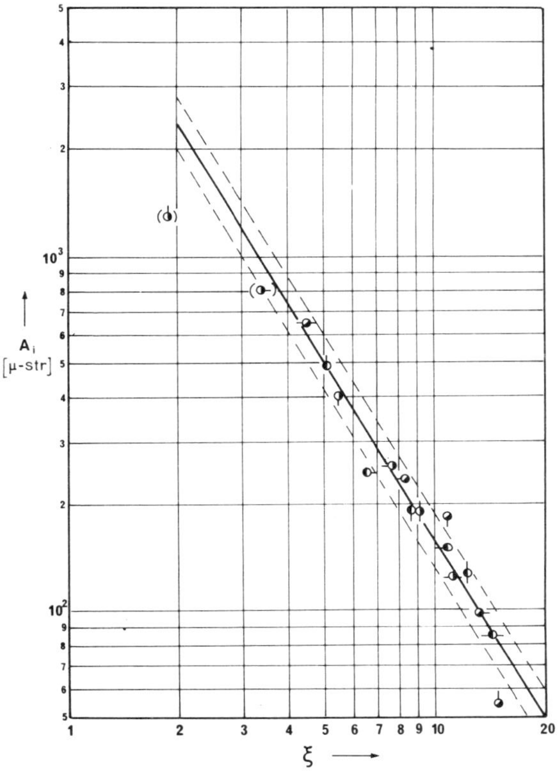

Fig. 90. Result of the evaluation of data from F57 (charge: 0.5 kg PV).

how the data points gather around a straight line. This line is determined by its slope, λ, and its value of A_0^* at $\xi = 1$ where:

$$\left.\begin{aligned} \lambda &= -1.68157 \\ A_0^* &= 7527\ \mu\text{-strain} \\ m &= \pm 17.6\% \end{aligned}\right| \quad [11.4]$$

The mean deviation, m, of the data points from this line is seen to be satisfactory.

Having completed this part of the evaluation, one must consider the dispersion data and the particle acceleration data. The non-dimensional distances for the pick-ups have already been determined, and these data must, therefore, correlate without any further correction of the distances. In order to be able to examine the results in relation to the theoretical results, the data must be non-dimensionalized properly.

Returning to 7.4, the non-dimensional quantities are to be expressed as follows:

$$\frac{\partial \sigma_r}{\partial t} = \frac{\partial S_r}{\partial t} \frac{p_0}{a} \sqrt{\frac{E_1}{\rho}} \qquad [11.5]$$

In the measurements the C_i data are given as:

$$C_i = \frac{\partial \epsilon_i}{\partial t} \qquad [11.6]$$

which by means of the calibration factor k_i(μ-strain/atm) for the pickups may be expressed as:

$$C_i = k_i \frac{\partial}{\partial t}\left(\frac{\epsilon_i}{k_i}\right) = k_i \frac{\partial \sigma_i}{\partial t} \qquad [11.7]$$

Making use of relation 7.24, these three equations will render:

$$\frac{\partial S}{\partial \tau} = \frac{aC_i}{p_0 c_s k_i} \qquad [11.8]$$

where σ_i is the value at pickup i for σ_r. One has thus obtained the normalizing factor, $a/p_0 c_s k_i$, for the C_i data.

One may now go one step further, and consider the particle acceleration, $\partial^2 v_r / \partial t^2$, which according to 7.4 is given in non-dimensional form as:

$$\frac{\partial^2 v_r}{\partial t^2} = \frac{\partial}{\partial \tau}\left(\frac{\partial v_r}{\partial t}\right) \cdot \frac{1}{a} \sqrt{\frac{E_1}{\rho}} \qquad [11.9]$$

If one now assumes, that locally, i.e. in the immediate neighborhood of the pick-ups, the conditions are such that the plane wave relations in 2.49 can be applied without serious discrepancies arising, one may rewrite 11.9 as follows:

$$\frac{\partial^2 v_r}{\partial t^2} = \frac{\partial}{\partial \tau}\left(\frac{\sigma_r}{E_1} c_s\right) \frac{1}{a} \sqrt{\frac{E_1}{\rho}} = \frac{\partial S}{\partial \tau} \frac{p_0}{E_1 a} \left(\sqrt{\frac{E_1}{\rho}}\right)^2$$

or:

$$\frac{\partial S}{\partial \tau} = \frac{\gamma a}{p_0 g} \frac{\partial^2 v_r}{\partial t^2} = \frac{\gamma a D_i}{p_0} \quad . \qquad [11.10]$$

It should be stressed that the D_i data represent the particle acceleration measured in g's. The normalizing factor, $\gamma a / p_0$, found in 11.10, rests on the assumption that the plane wave approximation used is permissible. It will be a matter of experimental evidence to prove or to disprove this assumption. This may be done by observing that according to 11.8 and 11.10:

$$\frac{aC_i}{p_0 c_s k_i} = \frac{\gamma a D_i}{p_0} \qquad [11.11]$$

where γ is the specific weight of the rock in its natural shape.

With this information, the data in Table XIII are now normalized according to 11.8 and 11.10 and given in Table XIV together with the non-dimensional distances. One will, however, observe, that the correct values of a and p_0 are not yet known. Because the attenuation curves appear as straight lines (theoretically) in a doubly logarithmic diagram, an error in the normalizing factor will only translate the straight line in the diagram vertically and parallel to itself. Thus its slope will remain unaffected. The normalizing factors have therefore been computed on the basis of the following data:

$$\left.\begin{array}{l} p_0 = 585.43 \text{ atm (anticipated)} \\ a = \text{those values already computed in Table XIII} \\ \gamma = 2.4 \text{ kg/dm}^3 \\ c_s = 3500 \text{ m/s (determined from the signals)} \\ k_i = k = 3.3982\ \mu\text{-strain/atm (common for all pickups)} \end{array}\right| \qquad [11.12]$$

One realizes that these values will have to be reconsidered at a later stage.

The data are plotted in Fig. 91 and it is observed how nicely they gather around straight lines as they should. This gives the first check on the validity of converting the charge factors, β, into properly adjusted values for the length a. The straight lines in the diagrams have slopes λ_a and λ_b for dispersion and particle accelerations, respectively. They have the respective values of C_0^* and D_0^* at $\xi = 1$, whereby:

$$\left.\begin{array}{ll} \lambda_a = -2.13285, & \lambda_b = -2.059697 \\ C_0^* = 15.31106, & D_0^* = 12.01210 \\ m = \pm 22.7\%, & m = \pm 6.4\% \end{array}\right| \qquad [11.13]$$

The mean deviation, m, of the data points from their respective lines indicate, that the uncertainty in the measurements of C_i is great, as indeed it is. Relation 11.11 is seen to have received its first experimental support. This may be observed already

TABLE XIV

Normalized C_i and D_i data from experiment F57 (0.5 kg PV)

Charge in	W(kg)	i	a/p_0c_sk (μs/μ-str)	ξ_i	$\frac{aC_i}{p_0c_sk}$	$\gamma a/p_0 \times 10^4$	$\frac{\gamma aD_i}{p_0}$
C	0.5	1	0.05758 496	5.113	0.61616	1.64375 6	
		2		6.609	0.25337		0.25313
		3		8.729	0.15548		
		4		10.974	0.13129		
		5		12.346	0.07198		0.06575
		6	(a = 0.40096)	14.466	0.03916		
		1	0.05731 353	1.879	4.12657	1.63600 8	
		2		3.383	0.81385		0.93252
		3		5.513	0.40406		
		4		7.768	0.20346		
		5		9.146	0.14099		0.14233
		6	(a = 0.39907)	11.267	0.07451		
		1	0.05226 967	14.837	0.03345	1.49203 1	
		2		13.189	0.06220		0.05565
		3		10.853	0.13381		
		4		8.380	0.20699		
		5		6.869	0.25612		0.23127
		6	(a = 0.36395)	4.534	0.52270		

in Table XIV. One would have liked to see $\lambda_a = \lambda_b$, which is almost the case in 11.13. In Fig. 92 the two straight lines of Fig. 91 have been plotted in one diagram, and the fact that these coincide within the accuracy of the measurements over the range where the data have been obtained is taken as a final support of 11.11 in this case.

The same procedure must be applied to the data obtained with the 0.625-kg Am charges. Table XV gives the data for A_i, C_i and D_i together with the distances, d_i, between the charge and the pickups. Again the mean value of the length a = 0.4 m has been assumed, the evaluation procedure (using the A_i data only) delivers the charge factors, β, as given in Table XV, whereupon the corrected values of a have been computed. Again the results have been normalized so that the mean position of the attenuation curve is determined by the shots in A and B only, because the charge factors of the shots in G and H indicate a much smaller effect of these shots. (Most probably this is caused by local irregularities in the rock.)

The result of this evaluation is shown in Fig. 93 where the data points gather nicely around a straight line when plotted as a function of the dimensionless distances, ξ, computed on the basis of the corrected values of a and given in Table XV. The attenuation curve (straight-line approximation) has the slope, λ, and the value of A_0^* at $\xi = 1$ given as:

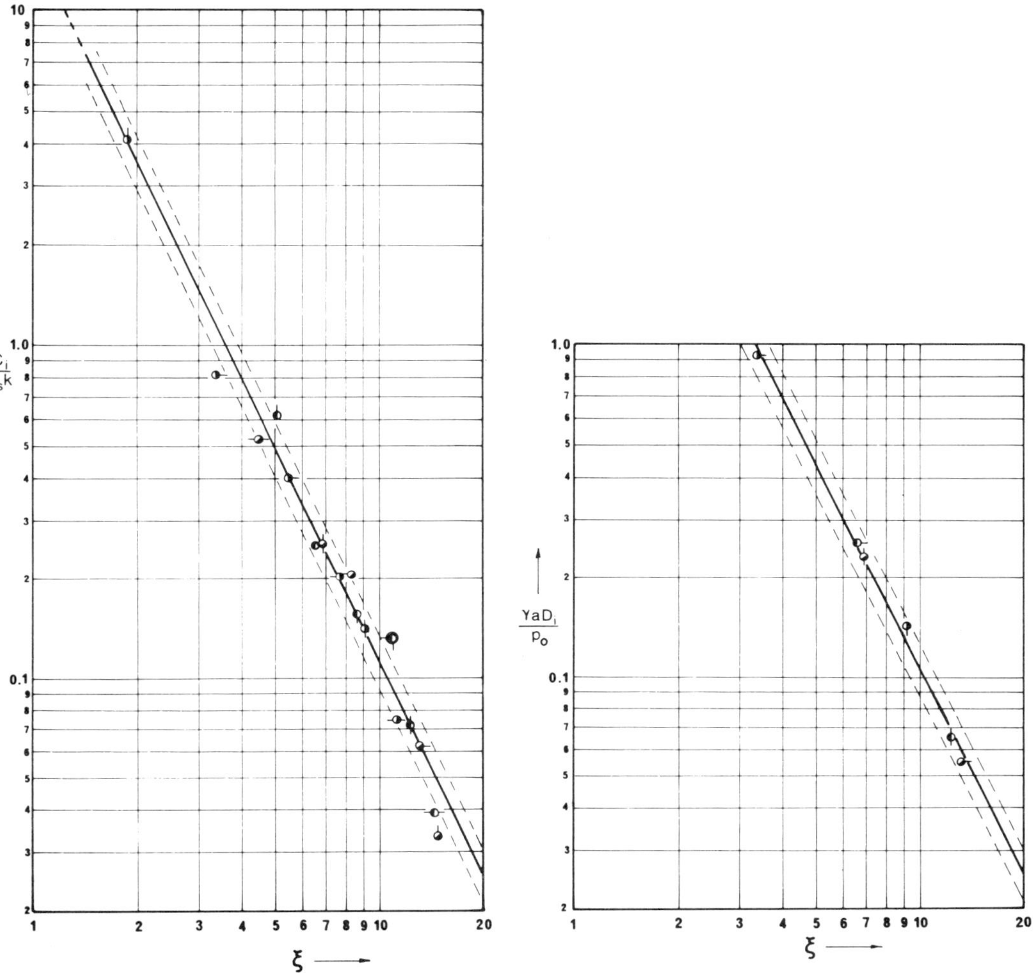

Fig. 91. Correlation of dispersion data (C_i) and particle acceleration data (D_i) by 0.5-kg charge (experiment F57).

$$\left.\begin{aligned} \lambda &= -1.56890 \\ A_0^* &= 5206\ \mu\text{-strain} \\ m &= \pm 7.4\% \end{aligned}\right| \qquad [11.14]$$

It is seen that the mean deviation, m, of the data point from the straight line is this time very satisfactory.

Now the remaining C_i and D_i data will have to be considered. The normalizing factors are this time computed with the following data:

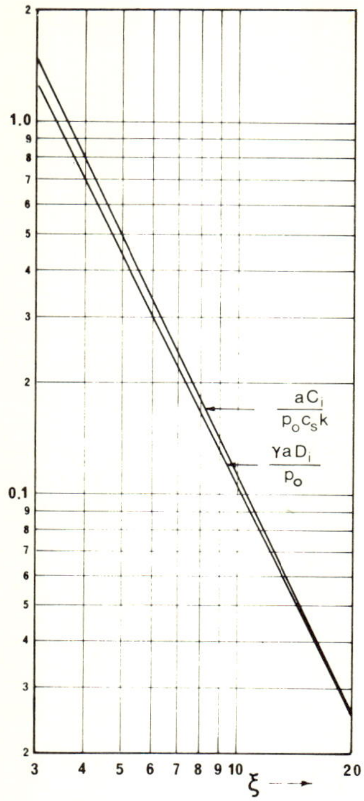

Fig. 92. Comparison between the attenuation curves for dispersion and particle acceleration (F57, 0.5 kg PV).

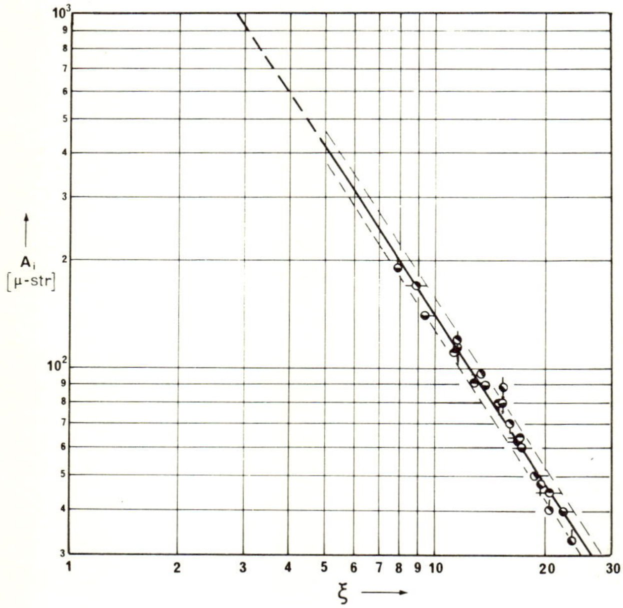

Fig. 93. Result of the evaluation of data from F57 (charge 0.625 kg Am).

TABLE XV

Experimental A_i, C_i and D_i data from experiment F57 (charge 0.625 kg Am)

Charge in	d_i(m)	A_i(μ-str)	D_i(m/s^2)	C_i(μ-str/μs)	β	a(m)	ξ_i
A	4.50	111		0.78			11.234
	5.10	91	(77)	0.605			12.732
	5.95	80		0.49	0.99777	0.40057	14.854
	6.85	64		0.31			17.101
	7.40			–			18.474
	8.25	45		0.27			20.596
B	3.15	190		1.64			7.886
	3.75	140	330	1.105			9.388
	4.60	115		0.81	1.00223	0.39943	11.516
	5.50	90		0.61			13.770
	6.05	80	170	0.63			15.147
	6.90	60		0.358			17.275
G	6.60	40		0.237			20.568
	6.00	50	90	0.245			18.690
	5.15	70		0.294	1.41300	0.32089	16.049
	4.25	96		0.350			13.244
	3.70	120	260	0.520			11.530
	2.85	170		0.523			8.882
H	8.05	33		0.405			23.776
	7.45	40		0.348			22.004
	6.60	48		0.537	1.29899	0.33857	19.494
	5.70	63		0.95			16.836
	5.15	88	150	1.06			15.211
	4.30	90		1.41			12.700

$$
\left.\begin{aligned}
p_0 &= 835.96 \text{ atm} \\
a &= \text{the values computed in Table XV} \\
\gamma &= 2.4 \text{ kg/dm}^3 \\
c_s &= 3500 \text{ m/s} \\
k_i &= k = 3.3982\ \mu\text{-strain/atm (common for all pickups)}
\end{aligned}\right| \qquad [11.15]
$$

and the resulting normalized data are obtained from the values of C_i and D_i in Table XV and given in Table XVI. When these are plotted against the already determined non-dimensional distances, the result will be as exhibited in Fig. 94. It is seen in Fig. 94a how nicely the dispersion data correlate with the non-dimensional distances. Fig. 94b shows that this is not the case for the particle acceleration

TABLE XVI

Normalized C_i and D_i data from experiment F57 (charge 0.625 kg Am)

Charge in	W(kg)	i	a/p_0c_sk (μs/μ-str)	ξ_i	$\frac{aC_i}{p_0c_sk}$	$\gamma a/p_0 \times 10^4$	$\frac{\gamma aD_i}{p_0g}$
A	0.625	1	0.04028 824	11.234	0.03142	1.15002	
		2		12.732	0.02437		0.00886
		3		14.854	0.01974		
		4		17.101	0.01249		
		5		18.747	–		
		6	(a = 0.40057)	20.596	0.01088		
B	0.625	1	0.04017 358	7.886	0.06588	1.14675	
		2		9.388	0.04439		0.03784
		3		11.516	0.03254		
		4		13.770	0.02451		
		5		15.147	0.02531		0.01949
		6	(a = 0.39943)	17.275	0.01438		
H	0.625	1	0.03227 424	20.568	0.00765	0.92126	
		2		18.690	0.00791		0.00829
		3		16.049	0.00949		
		4		13.244	0.01130		
		5		11.530	0.01678		0.02395
		6	(a = 0.32089)	8.882	0.01688		
G	0.625	1	0.03405 245	23.776	0.01379	0.97202	
		2		22.004	0.01829		
		3		19.494	0.01829		
		4		16.836	0.03235		
		5		15.211	0.03610	0.01458	
		6	(a=0.33857)	12.700	0.04801		

data and that at least one of the data points (in parenthesis) must be rejected. Linear regression gives the following values for λ_a and λ_b, and the values C_0^* of the straight lines at $\xi = 1$ for the dispersion and the particle acceleration data, respectively:

$$\lambda_a = -1.86525 \qquad \lambda_b = -2.01519$$
$$C_0^* = 3.03187 \qquad D^* = 3.55217 \qquad [11.16]$$
$$m = \pm 11.9\% \qquad m = \pm 15.6\%$$

Returning to 11.11 one would expect, for this relation to be supported, that $\lambda_a = \lambda_b$, which in this case is not confirmed. However, by plotting the two straight-line approximations to the attenuation curves in Fig. 94a and b in one diagram, as is done in Fig. 94c, one finds that even in this case [11.11] has received a reasonable

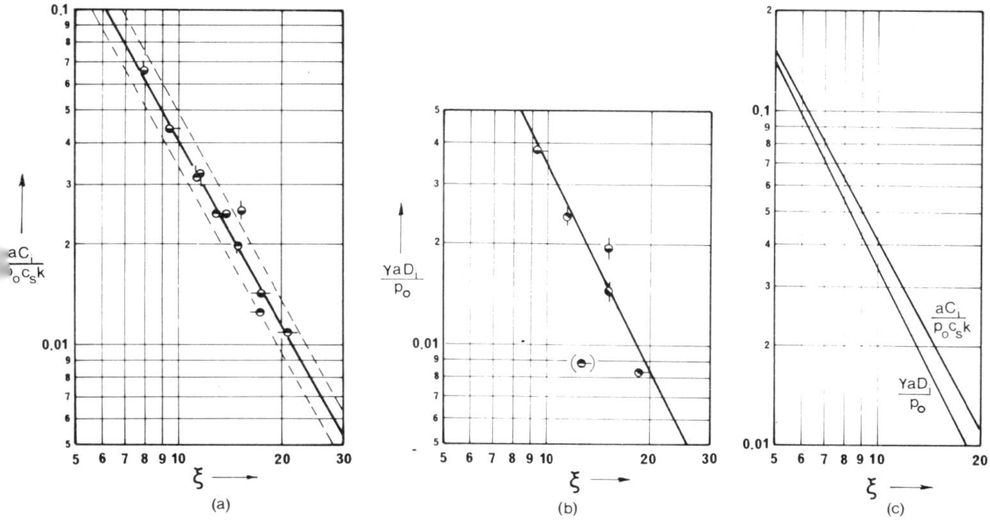

Fig. 94. Correlation of: (a) dispersion data, (b) particle acceleration data. (c) Comparison between the two attenuation curves. (F57, charge 0.625 kg Am).

support in view of the rather few and uncertain acceleration data. (Further remarks on the validity of 11.11 are found in Chapter 12.)

At this point in the evaluation the theoretical results must be drawn upon. The evaluation of experimental data has so far led to the determination of λ and λ_a. Because the validity of 11.11 is here assumed, and because, consequently, λ_a should be equivalent to λ_b as should also C_0^* to D_0^*, the values obtained are assumed interchangable, and the following set of results will be used in the following:

$$
\left.\begin{array}{lll}
\text{for 0.5 kg PV:} & \lambda = -1.68157 & \lambda_a = -2.05687 \\
 & A_0^* = 7527\ \mu\text{-strain} & C_0^* = 12.01210 \\
\text{for 0.625 kg PV:} & \lambda = -1.56890 & \lambda_a = -1.86525 \\
 & A_0^* = 5206\ \mu\text{-strain} & C_0^* = 3.03187
\end{array}\right| \qquad [11.17]
$$

The theoretical results predict the relationship between λ and λ_a on one hand and parameters κ and τ_0 of the problem on the other. This relationship is expressed in the diagrams in Figs. 69 and 87.

At this stage in the evaluation it is important to contemplate the situation. In contrast to what was previously the case, one has now, for each type of explosive, two pieces of information, namely λ and λ_a. It is therefore possible, again for each type of explosive, to determine two unknown parameters, namely κ and τ_0. Thus, first of all, the guesswork previously implied in determining τ_0 from a rather uncertain experimental value of t_0 and an estimated value of a is now replaced by a direct determination of the parameters, which will be shown subsequently. However, if the previous contentions are true, one may anticipate beforehand the results as follows:

(1) If the contention, that κ is a parameter describing the process, is true, one must expect the same κ-value for the two explosives, since the processes must be expected to be similar (with confined explosions in both cases).

(2) If the contention, that τ_0 is closely connected with the duration time of the input pulse, is true, it must be expected that the value of τ_0 for the two explosives must be different, with a higher τ_0-value for the explosive with the lowest combustion velocity (0.625 kg Am).

With these remarks in mind, the two diagrams in Figs. 69 and 87 are replotted in Fig. 95. The values of λ and λ_a in 11.17 are entered into their respective diagrams, and the combination of κ and τ_0, which in both diagrams simultaneously give the already established values of λ and λ_a, is determined. It is seen, that with a remarkable accuracy the κ-values obtained for the two explosives are indeed the same,

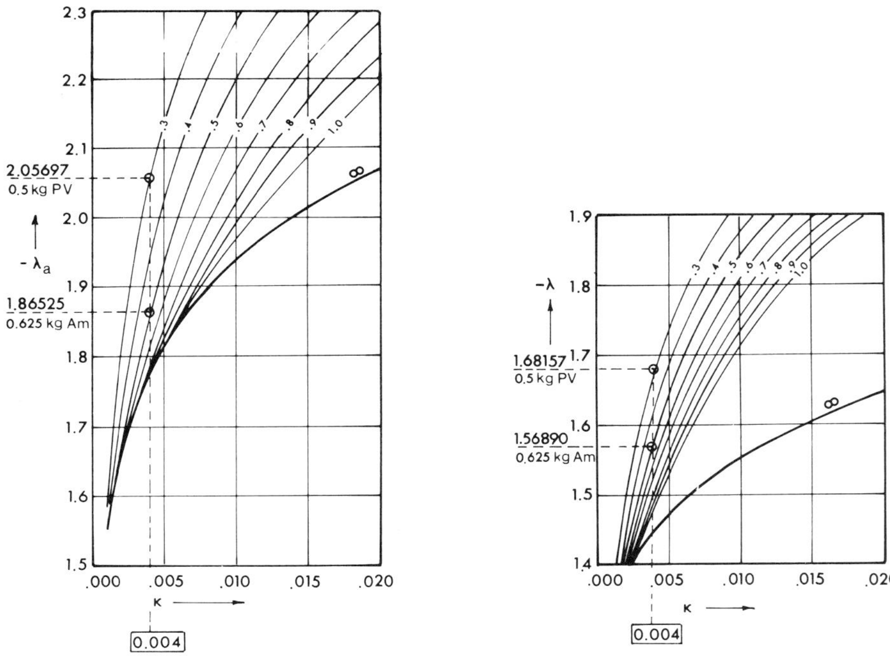

Fig. 95. Simultaneous determination of κ and τ_0 for F57 (charge: 0.5 kg PV and 0.625 kg Am).

whereas the τ_0 for 0.625 kg Am is greater than the corresponding τ_0-value for 0.5 kg PV. The result is:

$$\left.\begin{array}{lll} \text{for 0.5 kg PV:} & \kappa = 0.004, & \tau_0 = 0.3 \\ \text{for 0.625 kg Am:} & \kappa = 0.004, & \tau_0 = 0.5 \end{array}\right| \qquad [11.18]$$

This result must now be taken as a strong experimental support for the assumed applicability of the theoretical approach to the problem and the physical interpretations which have been given so far. It has been demonstrated, how the experimental data consisting of the maximum stress amplitudes, A_i, the dispersion data, C_i, and the particle acceleration data, D_i, give sufficient information to completely describe the wave propagation in rocks, by determining the necessary parameters in the theory.

Now one may return to the questions left open during the evaluation. These concern the normalizing factors and the values of length a used in the procedure. First it is realized that with κ and τ_0 known, it is now possible to find the appropriate values of p_0. The first step in this procedure is to utilize the formulae given in 9.1, 9.3, 9.4, 9.5, 11.1, 11.2 and 11.3 to compute the theoretical quantities given in Table XVII for the two cases. The computed values of λ and λ_a indicate, when compared with the values given in 11.17, the accuracy with which κ and τ_0 have been determined in the diagrams of Fig. 95.

The way in which this information is used is now displayed in Table XVIII. First it is noted, that the value of A_0^* at $\xi = 1$ of the straight-line approximation to the attenuation curve for the maximum strain amplitude is given in 11.4 and 11.14. These may be converted into stress (given in atm) by means of the calibration factor k whereby A_0 is obtained. Returning to Fig. 66 where the "overshoot" B is explained, one observes that the maximum pressure of the substitute rectangular input pulse may be computed as $p_0 = A_0/B$. Thus p_0 has been determined, but it is

TABLE XVII

Theoretical quantities

Theoretical quantity	0.5 kg PV	0.625 kg Am
λ	−1.67322	−1.57661
τ_*	0.18939	0.18939
β_∞	1.29158	1.29158
B	1.87532	2.14107
$\lambda_{a,\infty}$	−1.77540	−1.77540
λ_a	−2.05663	−1.86261
C_∞	2.65281	2.65281
ξ_0	4.70422	8.57017

TABLE XVIII

Correction procedure

Physical quantity		0.500 kg PV	0.625 kg Am
Value at $\xi = 1$ of the straight-line approximation to the attenuation curve (eqs. 11.4 and 11.11)	A_0^*	7527 μ-str	5206 μ-str
Calibration factor	k	3.3982 μ-st/atm	3.3982 μ-st/atm
A_0^* converted into stress: $A_0 = A_0^*/k$	A_0	2215.00 atm	1531.99 atm
Theoretical "overshoot" (Fig. 66)	B	1.87532	2.14107
Theoretical maximum value of the input pulse $p_0 = A_0/B$	p_0	1181.13 atm	715.53 atm
Normalizing factors (used in Table XIV and XVI) for C_i	$1/p_0 c_s k$	0.14362 μs/mμ-str	0.10058 μs/mμ-str
Corrected normalizing factors which shoild have been used	$1/p_0 c_s k$	0.07118 μs/mμ-str	0.11750 μs/mμ-str
Ratio f between the normalizing factors above	f	0.49565 247	1.16830 881
Value at $\xi = 1$ of the straight-line approximation to the attenuation curves (eq. 11.11)	C_0^*	12.01210	3.03187
Corrected value of C_0^*: $C^* = fC_0^*$	C^*	5.95383	3.54216
Theoretical value	C_0	4.10043	3.19943

stressed that this value is still influenced by the particular choice of a "mean value" of the length $a = 0.4$ m.

With this new value of p_0, the normalizing factors used in Tables XIV and XVI when considering the measured values of C_i can now be corrected as shown in Table XVIII. The result is that one obtains the ratio, f, between the normalizing factors used and the ones which should have been used had the correct value of p_0 been known at the time. The corrected values, C^*, are then easily computed, and may be compared with the theoretical value C_0. The latter one is found by returning to Fig. 86, from which it is realized that with the values of $\lambda_{a,\infty}$, d_a, C_∞ and ξ_0 given in Table XVII it is easy to construct the theoretical attenuation curves for $(\partial S/\partial \tau)_{max}$. These are shown in Fig. 96. Realizing that the experimentally obtained curves are now (after correction for p_0) given by:

$$\left(\frac{\partial S}{\partial \tau}\right)_{max} = C^* \xi^{\lambda_a} \qquad [11.19]$$

with C^* and λ_a given in the Tables XVIII and XVII respectively, one may plot these lines into the diagrams of Fig. 96 (dash-dotted curves). It is seen, that even after having been normalized properly, the theoretical and the experimental lines do not coincide.

The only remaining quantity, which has not been corrected and which still has

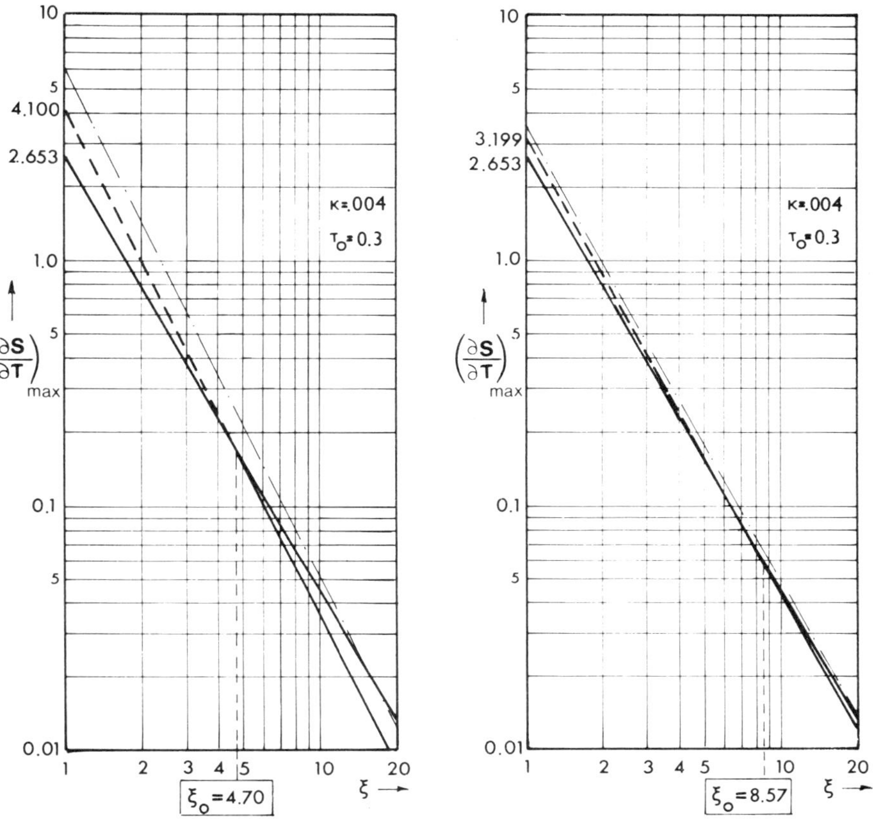

Fig. 96. Theoretical attenuation curves for $(\partial S/\partial \tau)_{max}$ for the two cases in F57 compared with the experimental ones (dash-dotted curves).

its assumed value is length a. The importance of a particular choice for a, which was made during the evaluation procedure, is exhibited in Fig. 97. The theoretical attenuation curve is shown with a = 0.4 m assumed, giving p_0 = 1181 atm in accordance with the result in Table XVIII for 0.5 kg PV. If now this length had been chosen a = 0.3 m, the result would have been that all non-dimensional distances would have been increased. The experimentally obtained curve would thus have been shifted horizontally (its slope remaining unchanged) to the position indicated in Fig. 97. The pressure would then have been p_0 = 1916 atm. The importance of this is brought out by realizing that the ratio a/p_0 appears in the normalizing factors in the Tables XIII and XV, and that in the present case this ratio has changed as follows:

$$\begin{aligned} &\text{for } a = 0.4: \quad a/p_0 = 0.00033\,87 \text{ m/atm} \\ &\text{for } a = 0.3: \quad a/p_0 = 0.00015\,66 \text{ m/atm} \end{aligned} \qquad [11.20]$$

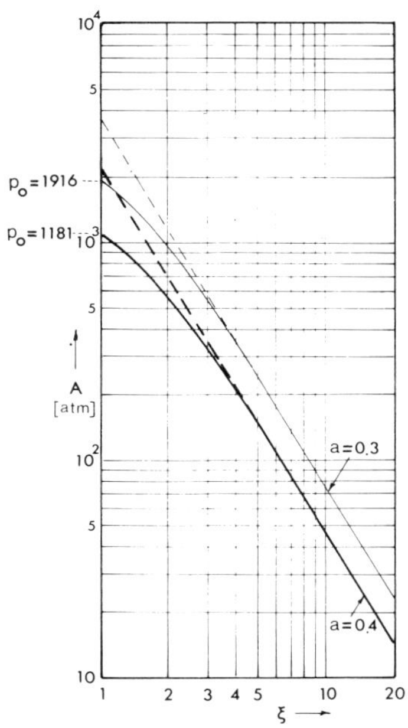

Fig. 97. The influence on p_0 of a change in the length a from a = 0.4 m to a = 0.3 m.

This means that the normalizing factor would thereby have been reduced by a factor of 0.46229. This means, however, that it is only a routine procedure to locate that value of a which would reduce the normalizing factors used so that C^* and C_0 in Table XVIII coincide. This determines the value of a in the present cases:

for 0.5 kg PV: $a = 0.348$ m
for 0.625 kg Am: $a = 0.3845$ m [11.21]

whereby theoretical and experimental results are brought in complete agreement.

There is a final check that should be mentioned. The duration time, t_0, of the positive phase of the signals (closest to the charge) were all of the order of 100 μs. Recalling the relation 9.9, it is now possible under the assumptions made, to estimate the "real" t_0, because τ_0 and a are now known. This gives:

for 0.5 kg PV: $t_0 = 59.7\ \mu s$
for 0.625 kg Am: $t_0 = 109.9\ \mu s$ [11.22]

which in view of the uncertainties in the measured values of t_0 is thought of as a very good agreement.

The present experiment has not only demonstrated the applicability of the theory and given it a nice experimental support, but it has also lent credence to the physical interpretations made based on this theory.

Chapter 12

SURFACE EXPLOSIONS

INTRODUCTION

So far the study of stress waves in rocks in this text has been limited only to waves created by confined explosions. It is, however, just as important to be able to describe stress waves created by explosions occurring on the surface. Such cases may arise if, for instance, a storage of explosives should accidentally be discharged, not to mention more disastrous cases.

Surface explosions will create craters of different sizes according to the magnitude, W, of the charge, and to the characteristics of the rock. From physical considerations one may conclude that the same charge W will create higher maximum amplitude of the stress wave when detonated confined than when detonated on the surface (unconfined). Crater building and energy transmission by surface explosions are thus questions which have been the prime target of investigation over a long period. Unfortunately this field of research is so wide and the results so uncertain that it is far beyond the scope of the present text to study it in detail. In this introductory course it has therefore been thought sufficient to give some results from a field experiment which may throw some light on the problem without pretending to provide final answers.

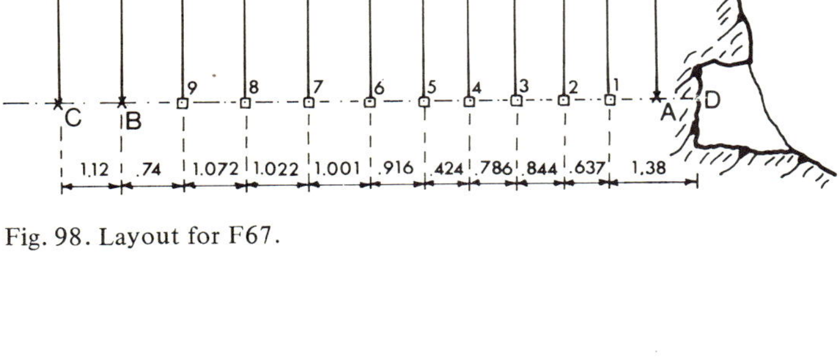

Fig. 98. Layout for F67.

EXPERIMENTAL LAYOUT

Experiment F67 which is discussed here took place at a location where a relatively steep precipice was found (Fig. 98). A short tunnel was excavated so as to get to that part of the rock which was untouched (more or less) by the erosion forces. The end of the tunnel was then used as the "free" surface and holes were drilled from the top of the precipice down to the level D, in which ordinary strain-measuring pickups were grouted. Holes C and B were used for detonation of charges of increasing magnitude. In this way the response to confined charges could be obtained with stress waves travelling in both directions. This could then be compared with the response from surface detonations at D. One obtained in this way also a check on possible pickup defects.

TABLE XIX

A_i and C_i data from experiment F67 (confined charges)

Charge in	W (kg)	i	A_i (μ-str)	k_i (μ-str/atm)	A_i (atm)	C_i (μ-str/μs)
C	0.5	4	24	3.5601	6.74	0.30
◑		6	41	3.7938	10.81	1.16
		7	62	3.7568	16.50	1.32
		8	132	3.4599	38.15	3.45
		9	(289)	4.1307	(69.96)	(8.93)
C	2.5	1	48	4.1857	11.47	0.259
●		2	43	4.8882	8.80	0.52
		4	75	3.5601	21.07	1.21
		5	101	3.4784	29.04	1.91
		6	144	3.7938	37.96	2.51
		8	(293)	3.4599	(84.68)	(4.85)
		9	(648)	4.1307	(156.87)	(11.99)
B	0.5	1	32	4.1857	7.65	0.73
◕		2	29	4.8882	5.93	0.63
		3	59	4.4564	13.24	1.82
		4	74	3.5601	20.79	2.89
		5	96	3.4784	27.60	2.96
		6	153	3.7938	40.33	5.23
		8	(429)	3.4599	(123.99)	(14.58)
		9	(1534)	4.1307	(371.37)	(57.95)
B	2.5	1	87	4.1857	20.79	1.14
◔		2	92	4.8882	18.82	1.27
		3	198?	4.4564	44.43?	2.86?
		4	158	3.5601	44.38	2.98
		5	209	3.4784	60.09	3.86
		6	326	3.7938	85.93	4.53
		8	(897)	3.4599	(259.26)	(14.08)
		9	(2262)	4.1307	(547.61)	(58.72)

THE CONFINED CHARGES OF F67

Data from the confined charges in C and B are given in Table XIX. They consist of the maximum strain values, A_i, and the dispersion data, C_i, as defined previously. A_i is measured in μ-strain and by means of the calibration factors k_i (which in this experiment varied so much that a mean value could not be used), each A_i is converted into pressure measured in atmospheres. Data appearing in parentheses were obtained so close to the charges, that they had to be disregarded in the subsequent evaluation.

The charges used were W = 0.5 kg PV and 2.5 kg PV in holes C and B. It is realized that pickup 3 exhibits erratic behavior after the sixth detonation (these data have a question mark) so all subsequent data from this pickup are neglected. A

TABLE XX

Evaluation of the A_i data (F67, confined charges)

Charge in	W (kg)	d_i (m)	a (theor.)	ξ_i (theor.)	β	a (corr.)	ξ_i (corr.)	A_i (atm)	i
C	0.5	6.29		20.98	1.14322	0.282	22.28	6.74	4
		4.95		16.51			17.54	10.81	6
		3.95	0.300	13.18			14.00	16.50	7
		2.93		9.77			10.38	38.15	8
		1.86		6.20			6.58	(69.96)	9
C	2.5	8.56		16.69	1.07347	0.497	17.23	11.47	1
		7.92		15.45			15.95	8.80	2
		6.29		12.27			12.67	21.07	4
		5.87	0.513	11.44			11.81	29.04	5
		4.95		9.66			9.97	37.96	6
		2.93		5.71			5.90	(84.68)	8
		1.86		3.62			3.74	(156.87)	9
B	0.5	7.44		24.81					1
		6.81		22.69					2
		5.96		19.88					3
		5.18	0.300	17.26					4
		4.75		15.84					5
		3.84		12.79					6
		1.81		6.05					8
		0.74		2.47					9
B	2.5	7.44		14.51	0.81486	0.563	13.23	20.79	1
		6.81		13.27			12.10	18.82	2
		5.96		11.62			10.60	44.43?	3
		5.18	0.513	10.09			9.20	44.38	4
		4.75		9.27			8.45	60.09	5
		3.84		7.48			6.82	85.93	6
		1.81		3.54			3.23	(259.26)	8
		0.74		1.45			1.32	(547.61)	9

further examination of the results shows that the amplifiers had been wrongly adjusted during the shot in B with the 0.5-kg charge. Finally, pickup 2 with its rather high value of its calibration factor, is suspected of giving wrong readings. This leaves only the data in Table XX to be considered in the evaluation procedure.

This detailed account of the pre-evaluation examination of the data has been mentioned to give an idea of the cautiousness one must exhibit when experimental data are to be evaluated.

Table XX exhibits now the evaluation procedure. First length a must be estimated, and in view of other results, $a = 0.3$ m for 0.5 kg PV has been assumed. Realizing that here charges of different magnitudes are to be evaluated together, the theoretical values for a for charge magnitudes, W, other than 0.5 kg is computed on the assumed validity of 10.3. In this way the theoretical a-values in Table XX are established. With these values the dimensionless distances ξ_i (theor.) are computed from the dimensional distances, d_i, given in the table. The standard evaluation procedure described in Chapter 8 delivers the charge factors, β, from which the corrected values a (corr.) and ξ_i (corr.) are computed. The final result is shown in Fig. 99 from which it is seen how nicely the data point gather around a straight line with the slope λ and a value of A_0^* at $\xi = 1$ given by:

$$\left.\begin{aligned} \lambda &= -2.22131 \\ A_0^* &= 6388.6 \text{ atm} \\ m &= \pm 6.0\% \end{aligned}\right| \qquad [12.1]$$

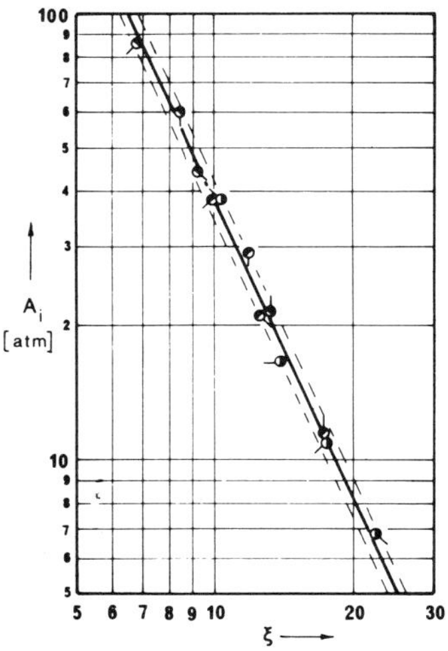

Fig. 99. Result of the evaluation for the A_i data in F67 (confined charges).

TABLE XXI

Legend

Point:	D	D	D	C	C	B	B	D	
Shot no.:	1	2	3	4	5	6	7	8	
Charge	0.5	1.0	2.0	0.5	2.5	0.5	2.5	5.0	
Legend	[illegible]	[illegible]	[illegible]	[illegible]	[illegible]	[illegible]	[illegible]	[illegible]	
Pickup no.	1	2	3	4	5	6	7	8	9
Legend	[illegible]	[illegible]	[illegible]	[illegible]	[illegible]	[illegible]	[illegible]	[illegible]	[illegible]

TABLE XXII

Normalized C_i data for experiment F67 (confined charges)

Charge	W (kg)	$a/A_0^*c_s \times 10^3$ (μs/atm)	k_i (μ-str/atm)	ξ	$\dfrac{aC_i}{A_0^*c_sk_i}$	i
4	0.5	7.63159	3.5601	22.28	0.00064 31	4
in C			3.7938	17.54	0.00233 34	6
([illegible])			3.7568	14.00	0.00268 15	7
			3.4599	10.38	0.00760 98	8
			4.1307	6.58	(0.01649 84)	9
5	2.5	13.45000	4.1857	17.23	0.00083 23	1
in C			4.8882	15.95	0.00143 08	2
([illegible])			3.5601	12.67	0.00457 14	4
			3.4784	11.81	0.00738 54	5
			3.7938	9.97	0.00889 86	6
			3.4599	5.90	(0.01885 40)	8
			4.1307	3.74	(0.03904 07)	9
6	0.5		4.1857			1
in B			4.8882			2
([illegible])			4.4564			3
			3.5601			4
			3.4784			5
			3.7938			6
			3.4599			8
			4.1307			9
7	2.5	15.23612	4.1857	13.23	0.00414 96	1
in B			4.8882	12.10	0.00395 85	2
([illegible])			4.4564	10.60	0.00977 81?	3
			3.5601	9.20	0.01275 35	4
			3.4784	8.45	0.01690 76	5
			3.7938	6.82	0.01819 27	6
			3.4599	3.23	(0.06200 31)	8
			4.1307	1.32	(0.21658 92)	9

The mean deviation, m, of the data points from this line is seen to be very satisfactory. The legend used in Fig. 99 and the subsequent figures of this chapter is shown in Table XXI.

The next step in the evaluation of the data is to normalize the C_i data according to 11.8. As was the case before, the exact values of a and p_0 are not yet known, but the normalizing factors are here computed based on the following set of data:

$$\left.\begin{aligned} &p_0 = 6388.6 \text{ atm } (= A_0^*) \\ &a = \text{the corrected values in Table XX} \\ &c_s = 5784 \text{ m/s} \\ &k_i = \text{the values given in Table XIX} \end{aligned}\right| \qquad [12.2]$$

Thus one obtains the normalized data for C_i given in Table XXII. Correlated with the corrected values ξ_i from Table XX these points are plotted in Fig. 100. They are seen to gather nicely around a straight line with slope λ_a and value of C_0^* at $\xi = 1$ whereby:

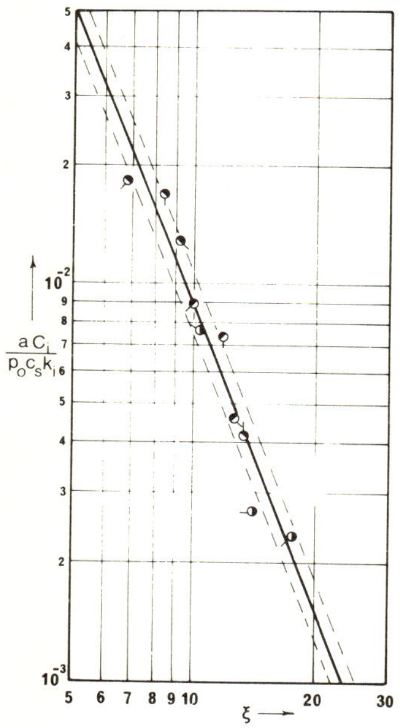

Fig. 100. The normalized dispersion data, C_i, as function of ξ (confined charges).

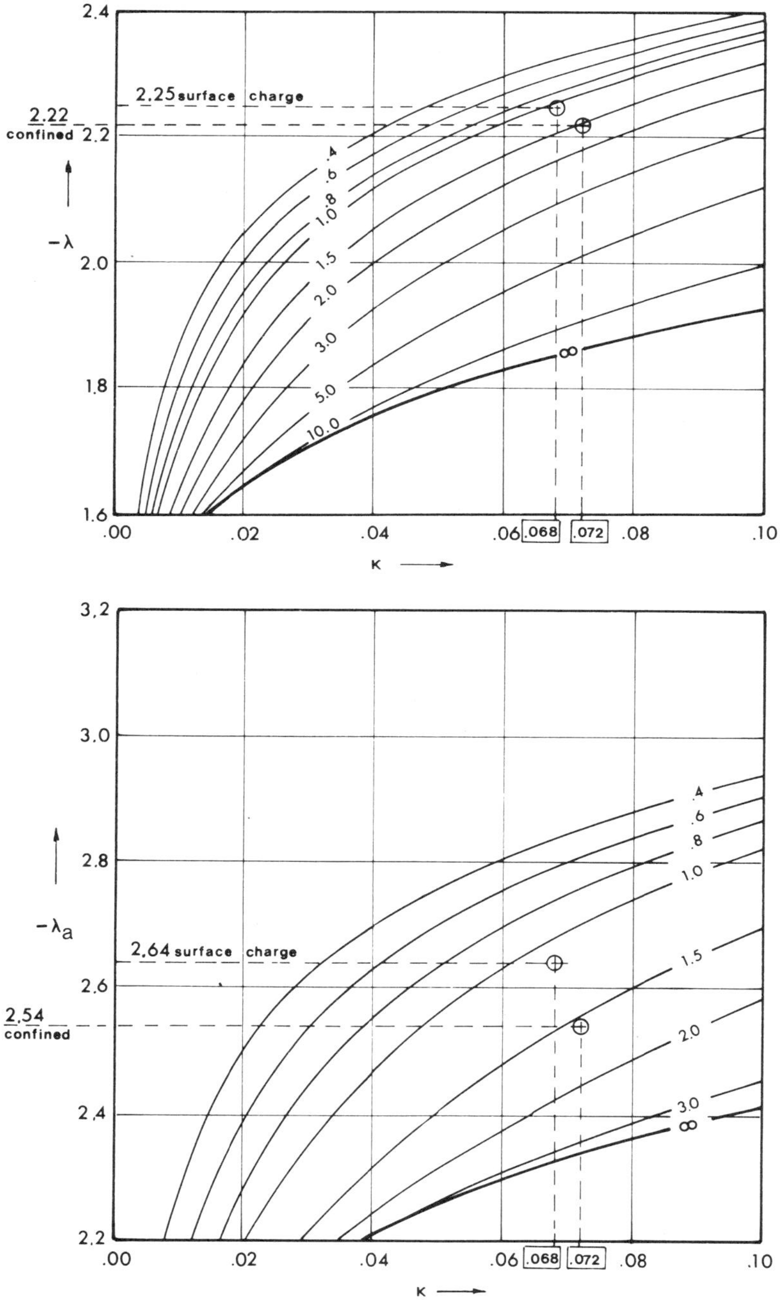

Fig. 101. Determination of κ and τ_0 from known values of λ and λ_a.

$$\lambda_a = -2.53890$$

$$C_0^* = 3.04641 \qquad [12.3]$$

$$m = \pm 19.8\%$$

The mean deviation, m, of the data points from this line is satisfactory in view of the uncertainties involved in their measurements.

One is now at the stage where parameters κ and τ_0 of the problem may be determined from the diagrams in Figs. 69 and 87. This is done in Fig. 101 with the result:

$$\kappa = 0.072 \qquad \tau_0 = 1.57 \qquad [12.4]$$

The further procedure follows that which has been given in detail in Chapter 11. This part is here postponed until the results of the surface charges have been considered.

THE SURFACE EXPLOSION

In the case of the surface explosions the evaluation may at first seem rather complicated. First, the theoretical solution for such cases has not been worked out. Thus the basic equations relating the important parameters are missing. It is of course a matter of routine to work out this solution, but certain features may be obtained without further analytic work.

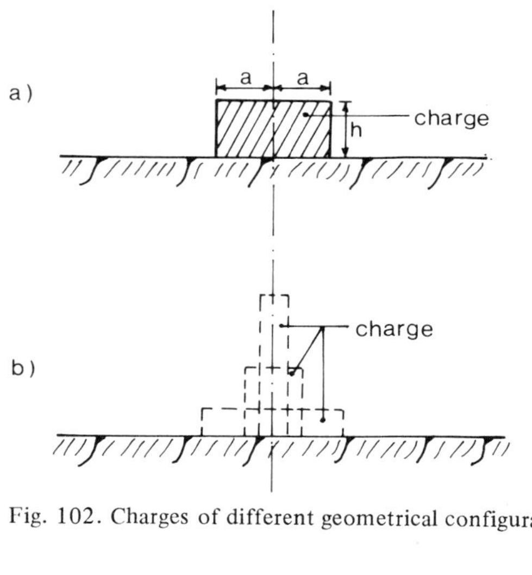

Fig. 102. Charges of different geometrical configurations on the surface of an infinite halfspace.

Consider the case of a charge on the surface of an infinite halfspace as indicated in Fig. 102a. First it is observed that apart from the geometrical dimensions of the load, the problem does not as such contain a length with which distances can be made dimensionless. It is thus clear that the stress waves created by a charge of a magnitude W will be different according to whether the charge is spread out over a large area or concentrated over a small one as indicated in Fig. 102b. This means that the weight, W, of the charge alone is not a sufficient description of the input, nor is the area which it initially covers.

The next consideration to be taken account of is the fact that a surface explosion will usually leave a crater, i.e. the situation after the explosion will be as sketched in Fig. 103. The crater radius, r_c, and the crater depth, h_c, therefore present themselves as lengths connected with the process which seem to be more suitable than the dimensions a and h of the charge in Fig. 102a.

The few remarks given here already give an idea of the complexity of the surface explosions, and it seems reasonable to try to find some empirical relations which can be used with some reliability to simplify the situation. One such empirical relation is the observation that for a large range of variation in W, the ratio between crater radius r_c and crater depth h_c is a constant:

$$\frac{h_c}{r_c} = 0.4 \qquad [12.5]$$

In this way one is left with a "theoretical" crater as sketched in Fig. 104. The volume of the masses "excavated" by the explosion can be computed through an assumed parabolic shape of the crater and then it turns out that for geometrically similar charges, the crater volume, V_c, is proportional to the charge weight, W. This means that one has an equivalent half-sphere with the radius, a, determined so that its volume is V_c. This presents the same situation for the surface explosion as previously mentioned (eq. 10.3) for the confined explosion, i.e.:

$$a \sim W^{1/3}. \qquad [12.6]$$

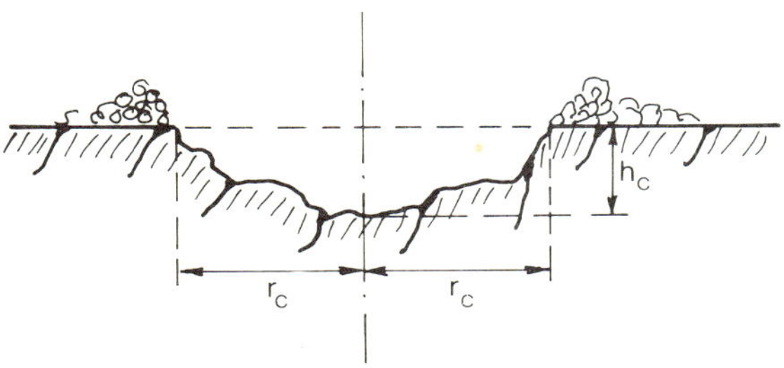

Fig. 103. Crater left by a surface explosion.

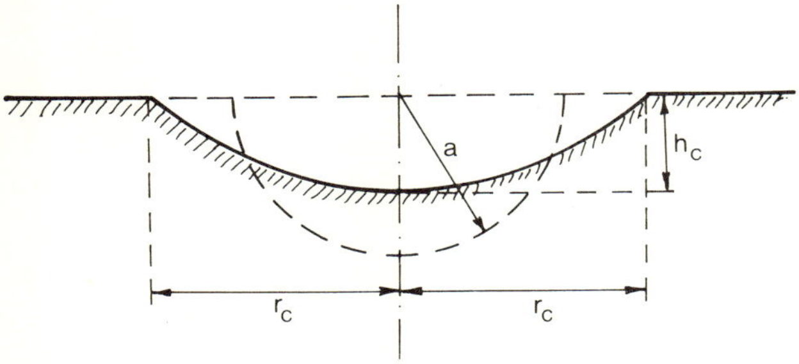

Fig. 104. The theoretical crater.

This radius a presents itself as a natural length in the problem with which the distances can be made dimensionless, and one has again replaced the input situation of the explosion with an "imaginary" sphere.

To bring out how arbitrary the choice of a is, attention is drawn to Fig. 105. Here the crater configuration is supposed to be spherical, introducing the radius, a, which in this case could be used to non-dimensionalize distances. This a would also satisfy 12.6, but one would have:

$$\left.\begin{array}{ll}\text{from Fig. 104:} & a = \frac{2}{3}\, r_c \\ \text{from Fig. 105:} & a = 1.45\, r_c\end{array}\right| \qquad [12.7]$$

Thus a variety of possible lengths to be used seems to present themselves, all of them, however, satisfying 12.6.

Here the results of these considerations may be contemplated further. If one considers the problem as it would present itself in an elastic medium, it is obvious

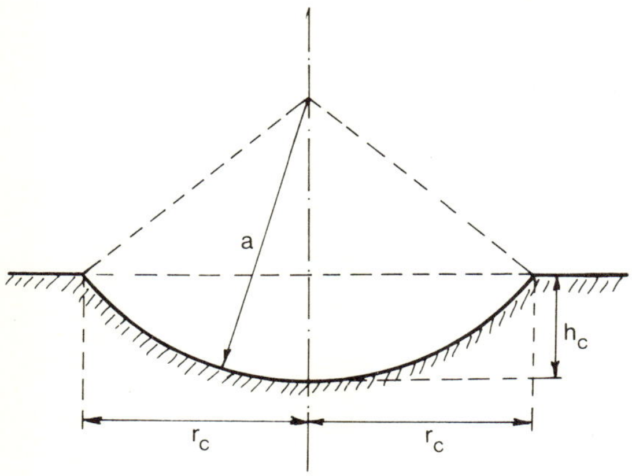

Fig. 105. The theoretical crater.

that a total theoretical description would consist of a compression wave followed by a shear wave due to the conditions at the free surface. Due to the fact that complete symmetry may be assumed around the vertical axis through the center of the crater, the shear wave would however, have to be zero at this axis. The situation along this axis would thus be similar to that of a confined explosion. This statement can be supported from considerations of the analytic solution. Thus it may be, that if measurements are made along this axis, the results from the solution to the confined explosion problem could still be applicable. This reasoning also holds true if the material is not elastic. Presentation of the experimental data from F67 may now give the answer.

TABLE XXIII

A_i and C_i data from experiment F67 (surface charges)

W (kg)	i	A_i (μ-str)	k_i (μ-str/atm.)	A_i (atm)	d_i (m)	C_i (μ-str/μs)
2.0	1	(2380)	4.1857	(568.6)	1.31	146.0
	2	780	4.8882	159.6	1.95	42.12
	3	370	4.4564	83.0	2.79	15.36
	4	136	3.5601	38.2	3.58	6.12
	5	118	3.4784	33.9	4.00	4.01
	6	93	3.7938	24.5	4.92	3.01
	7	40	3.7568	10.6	5.92	1.29
	9	26	4.1307	6.3	8.01	0.73
5.0	1	(3960	4.1857	(946.1)	1.24	130.4
	2	(1804)	4.8882	(369.1)	1.88	128.2
	3	738	4.4564	165.6	2.72	42.4
	4	336	3.5601	94.4	3.51	15.4
	5	282	3.4784	81.1	3.93	11.89
	6	208	3.7938	54.8	4.85	8.25
	7	88	3.7568	23.4	5.85	2.42
	8	59	3.4599	17.1	6.87	1.89
0.5	1	735	4.1857	175.6	1.38	50.0
	2	153	4.8882	31.3	2.02	8.67
	3	92	4.4564	20.6	2.86	3.67
	4	36	3.5601	10.1	3.65	–
1.0	1	(1340)	4.1857	(320.1)	1.34	75.93
	2	270	4.8882	55.2	1.98	14.12
	3	180	4.4564	40.4	2.82	7.20
	4	60	3.5601	16.9	3.61	1.63
	6	38.4	3.7938	10.1	4.95	
	7	18.7	3.7568	5.0	5.95	

THE SURFACE CHARGES OF F67

The surface charges, detonated at D in Fig. 98, consisted of steadily increasing charges placed in the small craters left by the preceding charge. W was increased as follows: $W_1 = 0.5$ kg PV, $W_2 = 1.0$ kg PV, $W_3 = 2.0$ kg PV and $W_4 = 5.0$ kg PV. The distances, d_i, from the center of the charges to the pickups are adjusted for the increasing crater depths, and d_i as well as the usual data A_i and C_i are given in Table XXIII. All these data are obtained on the normal to the free surface through the center of the charge, and one may now use these to test the validity of the previous considerations by attempting to apply the theoretical results for confined charges to these data.

The following remarks on the data should be kept in mind. By scrutinizing the

TABLE XXIV

Evaluation of the A_i data for experiment F67 (surface charges)

W (kg)	i	a (theor.)	ξ_i (theor.)	β	a (corr.)	ξ_i (corr.)	A_i (atm)
2.0	1	0.238	(5.50)	0.88980	0.251	(5.23)	(568.6)
	2		8.19			7.78	159.6
	3*		11.72			11.13	83.0
	4*		15.04			14.28	38.2
	5*		16.81			15.96	33.9
	6*		20.67			19.62	24.5
	7		24.87			23.61	10.6
	9		33.66			31.96	6.3
5.0	1	0.323	(3.84)	0.79260	0.358	(3.46)	(946.1)
	2		(5.82)			(5.25)	(396.1)
	3		8.42			7.59	165.6
	4*		10.87			9.80	94.4
	5*		12.17			10.97	81.1
	6*		15.02			13.54	54.8
	7		18.11			16.33	23.4
	8		21.27			19.18	17.1
	1	0.150	9.20	1.22613	0.137	10.07	175.6
	2		13.47			14.75	31.2
	3*		19.07			20.88	20.6
	4*		24.33			26.64	10.1
1.0	1	0.189	7.09	1.15642	0.177	7.56	320.1
	2		10.48			11.18	55.2
	3*		14.92			15.92	40.4
	4*		19.10			20.38	16.9
	6*		26.19			27.94	10.1
	7		31.48			33.58	5.0

data for systematic behavior, some doubt was placed on the positions (d_i) of pickups 8 and 9. Data from these pickups were therefore neglected in the evaluation procedure. The same holds also for pickup 2, which also in the case of the confined charges appeared to give dubious results. This leaves only the data which in Table XXIV have been denoted by a star to be considered in the evaluation procedure. These are the same ones which in the case of the confined charges gave significant data, and one has the additional assurance, that pickup differences will not affect the comparison which is the main purpose here.

The data evaluation starts with assuming a length a with which the distances can be made dimensionless. From the measured crater dimensions this length was assumed equal to the crater depth, $h_c = 0.15$ m $= a$ for 0.5 kg PV. Assuming furthermore the validity of 12.6, the appropriate values of $a = a$ (theor.) may be

TABLE XXV

Normalized C_i data for experiment F67 (surface charges)

Charge	W (kg)	$a/A_0^* c_s \times 10^{-3}$ (μs/atm)	k_i (μstr/atm)	ξ	$aC_i/p_0 c_s k_i$	i
3	2.0	2.47015	4.1875	(5.23)	0.08616	1
in D			4.8882	7.78	0.02128	2
			4.4564	11.13	0.00851	3
			3.5601	14.28	0.00425	4
			3.4784	15.96	0.00285	5
			3.7938	19.62	0.00196	6
			3.7568	23.61	0.00085	7
		(a = 0.251)	4.1307	31.96	0.00044	9
8	5.0	3.52316	4.1857	(3.46)	0.10976	1
in D			4.8882	(5.25)	0.09240	2
			4.4564	7.59	0.03352?	3
			3.5601	9.80	0.01524	4
			3.4784	10.97	0.01204	5
			3.7938	13.54	0.00766	6
			3.7568	16.33	0.00227	7
		(a = 0.358)	3.4599	19.18	0.00192	8
1	0.5	1.34825	4.1857	10.07	0.01611	1
in D			4.8882	14.75	0.00239	2
			4.4564	20.88	0.00111	3
		(a = 0.137)	3.5601	26.64		
2	1	1.74190	4.1857	7.56	0.03160	1
in D			4.8882	11.18	0.00503	2
			4.4564	15.92	0.00281	3
			3.5601	20.38	0.00080	4
			3.7938	27.94		6
		(a = 0.177)	3.7568	33.58		7

computed and the result is shown in Table XXIV with the whole evaluation procedure. The result is exhibited in Fig. 106 where the data points now as usual gather around a straight line with slope λ and a value A_0^* at $\xi = 1$ given as:

$$\lambda = -2.24798$$

$$A_0^* = 17568 \text{ atm} \qquad [12.8]$$

$$m = \pm 10.6\%$$

It is seen that the mean deviation, m, of the data points from the straight line is again very satisfactory.

The next step is to consider the dispersion data, C_i, which now will have to be made dimensionless according to 11.8 with following data:

$$p_0 = 17568 \text{ atm } (= A_0^* \text{ in 12.8})$$

$$a = \text{values computed in Table XXIV} \qquad [12.9]$$

$$c_s = 5784 \text{ m/s (determined from signals)}$$

$$k_i = \text{values in Table XXIII}$$

In this way the normalized data in Table XXV are obtained and these are correlated with the corrected values of ξ_i in Fig. 107. One will here find, that there is a tendency of systematic deviations between data from the different charges, which

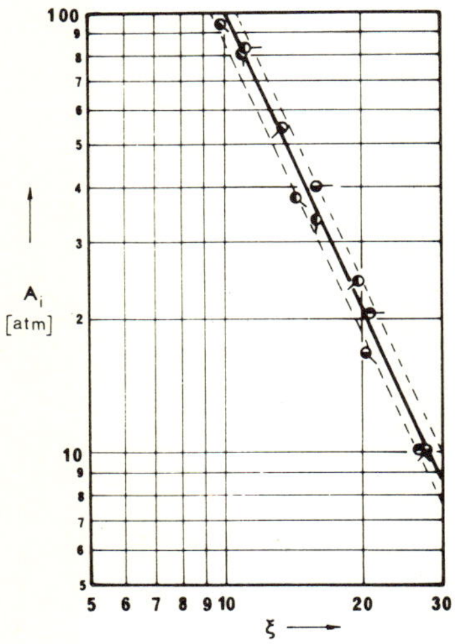

Fig. 106. Result of the evaluation of the A_i data for F67, surface charges.

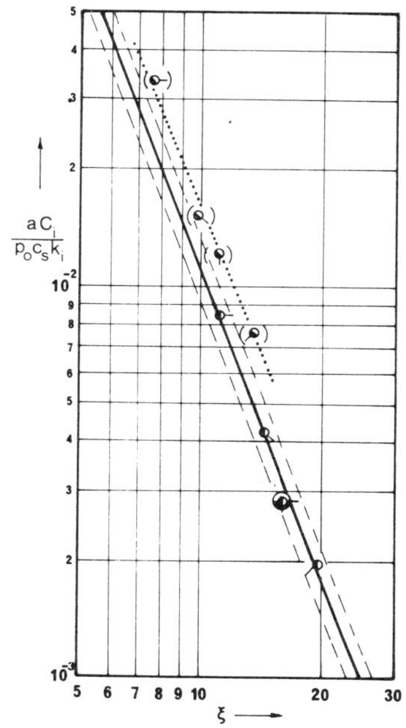

Fig. 107. Correlation between the normalized C_i data and ξ (surface charges).

presumably should have been already accounted for in the determination of the corrected values of ξ. This appears as a shift in the data as exhibited in Fig. 107, where the data points in parentheses lie on a straight line (dotted) parallel to the solid line. The slope, λ_a, of the latter one and its value of C_0^* at $\xi = 1$ is therefore determined from the data of charge 3 in D which gives:

$$\left.\begin{aligned} \lambda_a &= -2.64045 \\ C_0^* &= 4.75198 \\ m &= \pm 6.7\% \end{aligned}\right| \quad [12.10]$$

One has here established λ and λ_a experimentally for the surface charges, and may introduce these values into the diagrams of Fig. 101 to determine simultaneously the values of parameters κ and τ_0. This gives:

$$\kappa = 0.068, \qquad \tau_0 = 1.13 \qquad [12.11]$$

This procedure assumes the applicability of the theory for confined charges to the particular data from surface charges. It is expected from previous considerations that the κ-values should be the same and that the τ_0-values should be different in such a way that τ_0 by surface charges should be less than for confined charges. A comparison between the results in 12.4 and 12.11 shows that this is just the case. The two κ-values differ by an amount which is smaller than the accuracy with which they are determined. Before further conclusions are drawn, the particle acceleration data will be considered.

PARTICLE ACCELERATION DATA

In this case also the particle accelerations, D_i, were measured and one has therefore the possibility to check relation 11.11. One must, however, realize that the evaluation so far has shown systematic deviations in the data, which account for the neglection of data from charge no. 6 in the confined case and which showed up in the C_i data in Fig. 107. Similar behavior of the D_i-data must therefore be expected.

The D_i-data are given in Table XXVI. They are normalized according to 11.10, and the values in 12.9 are used in addition to which it is noted that the specific weight of the rock was $\gamma = 2.4$ kg/dm^3. Thus the normalized data in Table XXVI are obtained and correlated with the corrected values ξ_i of the non-dimensional distances. For the confined charges this correlation is shown in Fig. 108. It is noted that again a shift in the data is present so that some of the data gather around a straight line (shown with a dotted line) parallel to the solid line. The latter one has a slope λ_b and a value D_0^* at $\xi = 1$ given by:

$$\left.\begin{aligned} \lambda_b &= -2.41761 \\ D_0^* &= 4.57910 \\ m &= \pm 22.8\% \end{aligned}\right| \qquad [12.12]$$

This value of λ_b should be compared with λ_a in 12.3. In view of 11.11, these values ought to be the same. Due to the fact that the present data are somewhat uncertain, the agreement between the two values is considered satisfactory. The correlation of the normalized D_i data for the surface charges is shown in Fig. 109. These data exhibit the same shift as the ones observed earlier in Figs. 107 and 108. Thus the data from charge no. 2 and no. 3 indicate straight lines with slopes $\lambda_b = -2.73776$ and $\lambda_b = -2.80573$, respectively. These values should be compared with the value in 12.10 and again the agreement is fair. However, the check on 11.11 implies not only that the slopes should be the same, but the experimentally obtained lines should coincide. In Fig. 109 the lines from Fig. 107 have therefore been plotted for comparison. It is seen that this comparison is *not* good. This is of course due to the shift in the data, which is caused by the fact that the values of *a*

TABLE XXVI

Particle acceleration data (D_i) from experiment F67

Charge	i	D_i	$\gamma a D_i/p_0$	ξ_i	Charge	i	D_i	$\gamma a D_i/p_0$	ξ_i
Surface charges					*Confined charges*				
3	1	–			4	1	–	–	
at D	2	21792	0.07472 39	7.78	in C	2	–	–	
	3	7747	0.02656 42	11.13		3	141	0.00149 37	25.09
	4	2607	0.00893 93	14.28		4	250	0.00264 85	22.28
	5	2415	0.00828 09	15.96		5	385	0.00407 86	20.80
	7	868	0.00297 63	23.61		7	672	0.00711 91	14.00
	8	518	0.00177 62	27.66		8	1809	0.01916 43	10.38
	9	306	0.00104 93	31.96		9	4794	0.05078 70	6.58
8	1	–	–		5	1	–	–	
at D	2	–	–		in C	2	220	0.00410 76	15.95
	3	23808	0.11643 80	7.59		3	413	0.00771 10	14.24
	4	6916	0.03382 42	9.80		4	600	0.01120 25	12.67
	5	6506	0.03181 90	10.97		5	770	0.01437 65	11.81
	7	1458	0.00713 07	16.33		7	1045	0.01951 09	7.95
	8	1270	0.00621 12	19.18		8	2278	0.04253 20	5.90
	9	873	0.00426 96	22.18		9	6556	0.12240 55	3.74
1	1	20090	0.03760 01	10.07	6	1	–		
at D	2	4077	0.00763 05	14.75	in B	2	190		
	3	1764	0.00330 15	20.88		3	592		
	4	560	0.00104 81	26.64		4	1140		
	7	182	0.00034 06	43.73		5	1732		
2	1	–	–			7	3190		
at D	2	7575?	0.01831 66?	11.18		8	7236		
	3	3365	0.00813 67	15.92		9	42,300		
	4	1106	0.00267 43	20.38	7	3	768	0.01624 33	10.60
	5	1067	0.00258 00	22.79	in B	4	800	0.01692 01	9.20
	7	399	0.00096 48	33.58		7	3630	0.07677 51	5.03
	8	179	0.00043 28	39.39		8	7236	0.15304 27	3.23
	9	120	0.00029 01	45.45		9	36,660	(0.77536 54)	(1.32)

determined in the evaluation of the A_i data may not give the right relation between the a-values. This could to some extent be caused by errors in the determination of the positions of the pickups. It may, however, be concluded, that the slopes λ_a used in the previous section for the determination of parameters κ and τ_0 have been confirmed with sufficient accuracy by the D_i data. Further discussion is deferred to the next section.

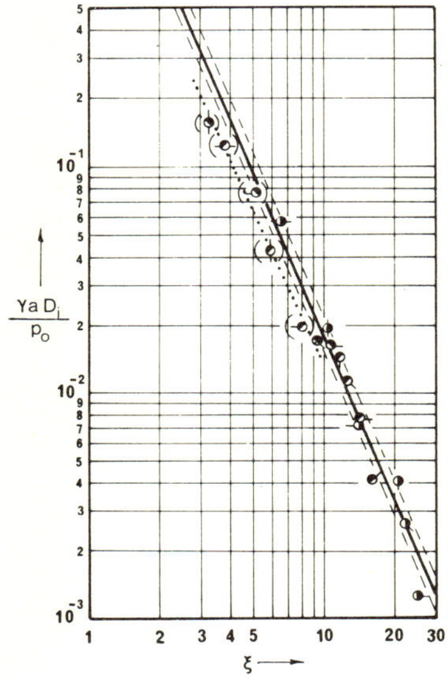

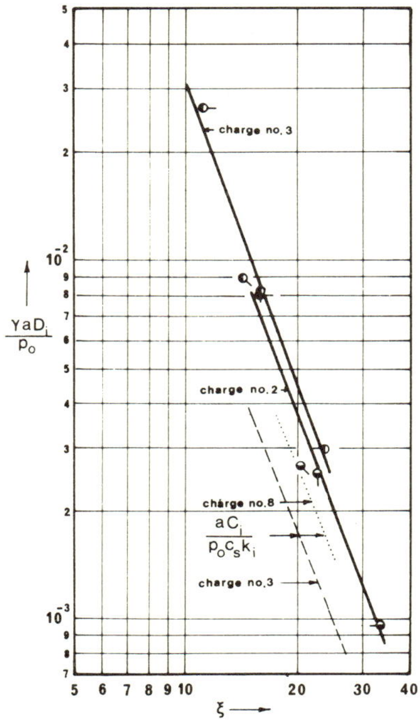

Fig. 108. Correlation of the D_i data for F67 (confined charges) exhibiting the shift in the data.

Fig. 109. Correlation of the D_i data showing the systematic shift (surface charge). The attenuation curves from Fig. 107 are shown for comparison.

FINAL CONCLUSIONS

The evaluation of the data from this experiment have revealed systematic errors in the data, causing the shifts pointed out in the previous discussions. Certain conclusions may, however, be drawn, which have a general character.

(1) In spite of the shifts appearing in the diagrams, slopes λ and λ_a are determined with sufficient accuracy in 12.1, 12.3, 12.8 and 12.10. It is noted that these values are the ones from which the parameters κ and τ_0 are determined.

(2) Based on theoretical considerations and the concept of the process of a detonating confined charge as compared with that of a detonating surface charge, it is suspected, that on the normal to the surface through the center of the surface charge, the two processes may be conceived of as being the same. This is supported by the fact, that one experimentally obtains the same κ-value for the two cases in Fig. 101.

(3) The first conclusion to be drawn is thus that one has found a way in which

to connect the two types of explosions and the stress wave propagation caused by them.

The next step must be to explore the consequences of such a conclusion. The previously established theoretical formulae in 9.1, 11.1, 11.2 and 11.3) may now be used to compute the quantities in Table XXVII. The values computed for λ and λ_a should be compared with the values in 12.1, 12.3, 12.8 and 12.10 to obtain an idea of the accuracy with which κ and τ_0 are determined in Fig. 101.

The further procedure follows as exhibited in detail in Chapter 11. The input pressure, p_0, is computed from A_0^* by means of the "overshoot", B. The intersection, C_0, of the straight line approximations to the attenuation curves for $(\partial S/\partial\tau)_{max}$ with $\xi = 1$ (see Fig. 86) may be computed from geometrical considerations, and the result is given in Table XXVI. The normalizing factors used in establishing C_0^* in 12.3 and 12.10 may then be corrected, and one obtains:

$$\begin{aligned} &\text{confined charges (assumed } a = 0.3 \text{ m):} \quad C_0 = 5.46109 \\ &\text{surface charges (assumed } a = 0.15 \text{ m):} \quad C_0 = 9.36717 \end{aligned} \qquad [12.13]$$

These values should correspond to the values in Table XXVI, which they do not. Thus the values of a are sought, which in the two cases will change the normalizing factors such that the correspondence is achieved. This will result in the following values for a:

$$\begin{aligned} &\text{confined charges:} \quad a = 0.2042 \text{ m} \\ &\text{surface charges:} \quad a = 0.0915 \text{ m} \end{aligned} \qquad [12.14]$$

TABLE XXVII

Computed constants for the two cases

	Confined charge	Surface charge
κ	0.072	0.068
τ_0	1.57	1.13
λ	−2.21890	−2.24425
B	1.79263	1.97121
$\lambda_{a,\infty}$	−2.34100	−2.32823
λ_a	−2.54047	−2.64183
C_∞	1.00645	1.03285
ξ_0	9.62236	6.72940
C_0	1.58099	1.87798

(It is stressed that these values refer to charges 0.5 kg PV.) This then concludes the information obtainable from the experiment.

However, one may pose the following question: What is the relation between the maximum stress amplitude created at a given distance by a surface charge as compared with that created by a confined charge of the same magnitude? This question can be answered by observing, that the straight-line approximation to the attenuation curves for the maximum stress amplitudes, A, in the two cases may now be expressed as:

$$\begin{aligned} &\text{confined charge:} \quad A = 15014\,\xi^{-2.22131}\ \text{atm} \quad (a = 0.2042) \\ &\text{surface charge:} \quad A = 53370\,\xi^{-2.24798}\ \text{atm} \quad (a = 0.0915) \end{aligned} \qquad [12.15]$$

where ξ is computed with the a-values indicated. At a certain distance, e.g. 5 m, the values of ξ will be different in the two cases, and one obtains:

$$\begin{aligned} &\text{confined charge:} \quad d = 5.0\ \text{m} \rightarrow \xi = 24.49 \rightarrow A = 12.34\ \text{atm} \\ &\text{surface charge:} \quad d = 5.0\ \text{m} \rightarrow \xi = 54.64 \rightarrow A = 6.33\ \text{atm} \end{aligned} \qquad [12.16]$$

The answer is accordingly, that the maximum stress amplitude is twice as great in the confined case as in the case of a surface charge. This holds true at other distances as well, because the slopes of the two curves are almost identical. It should be mentioned that this result is supported by other investigations for "half buried" charges.

The last question which remains unanswered is the relation between the proper value of a and the crater dimensions discussed previously in this chapter. The measurement of crater depths involves great uncertainties because, even in the same rock massive, great differences may be caused by local conditions. However, the indications of the present experiment are that the proper value of a for surface charges is:

$$h_c < a < \tfrac{5}{3}\, h_c \qquad [12.17]$$

i.e. a lies between the crater depth, h_c, and the radius of the crater volume, V_c (Fig. 104).

A final comment to this experiment is to mention that not too far-reaching conclusions should be drawn from it. The conclusions already drawn are those which have support from other sources and which have been verified beyond what is shown here by own experiments.

Chapter 13

SOME PRACTICAL PROBLEMS

INTRODUCTION

The present text would somehow seem incomplete if practical applications of the theory were not discussed. Remaining strictly within applications in rocks, one can conceive of a large variety of practical problems where a good understanding of the propagation of stress waves is more or less indispensable. For a long time people have excavated mines, with all the problems of keeping existing shafts safe when new ones were to be excavated. As our technological capability improves, the possibility of using rock sites for underground facilities like power plants, factories, etc., may become an attractive alternative. In any such development the theory of stress-wave propagation in rocks becomes an important tool.

It is not the scope of the present chapter to give an extensive survey of all such practical applications, but rather to select some essential problems which may serve as a guide when new challenges are to be met.

THE EXTRAPOLATION PROBLEM, 1

It will have been noticed that in most of the cases treated in this text, the standard row experiment has been drawn to attention whenever support for theoretical results has been desired. This is due to the fact that these experiments are the only ones which were not only readily available to the author, but which also delivered all the essential data and information for a successful interpretation. However, these experiments were in most cases done with 0.5–2.5-kg charges, and inevitably the question arises: Can the results be applied with confidence when the charges increase by many orders of magnitude? This is known as the extrapolation problem and it can only be answered by drawing upon data presented in the literature from large-scale experiments.

Fig. 110 represents replots of two line drawings exhibiting experimentally obtained attenuation curves for particle velocity and acceleration from large-scale explosions. The Cowboy experiment data were obtained in salt with charges ranging from 20 to 1000 lbs. of pelletized TNT. The data in salt, granite and tuff were obtained from nuclear charges. The two diagrams represent an attempt to gather data from experiments with different explosives and different charge magnitudes in different types of rocks in one and the same diagram. The difference in charge magnitudes is accounted for by absorbing the effect in the normalizing of the coordinates. The difference in explosives is accounted for by normalizing the charge to TNT-equivalence and the difference in the rock type is taken care of by scaling factors or "coefficients". The philosophy behind this "scaling" procedure touches upon many questions of a fundamental nature, some of them will be commented upon here and in the next section.

First, it is noticed that the abscissa used in Fig. 110 is called "the slant range". It is not a non-dimensional distance, but it carries in it the feature of scaling distances, d, with $W^{1/3}$ as was exhibited in the conversion of the charge factors, β, into appropriate values of the radius, a. Under certain ideal conditions this works very well (see Chapter 10, F46, shots in F) but most frequently this is only a fair approximation and will have to be adjusted (see Chapter 10, F46, shots in E).

The question raised here is now answered by plotting the results from a standard row experiment into the diagrams of Fig. 110. The experiment F57, which has been

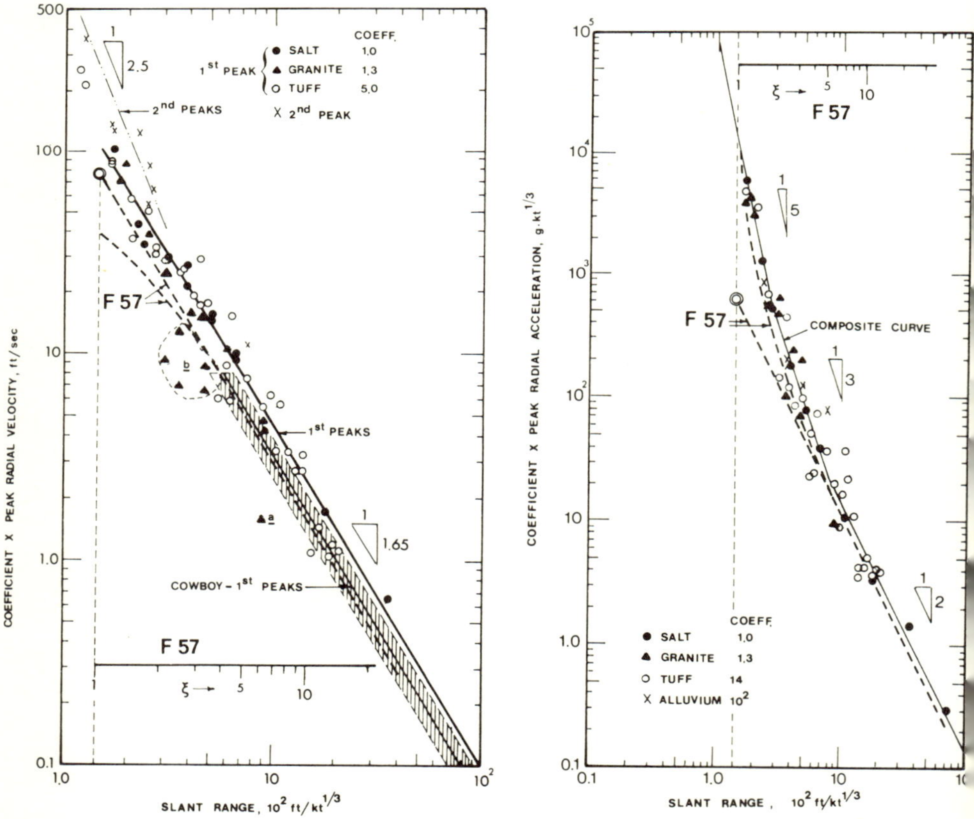

(a) (b)

Fig. 110. Replot of data from large-scale experiments with the results of F57 drawn for comparison (0.5 kg PV).

treated in detail in Chapter 11, has been chosen as an example, and it is stressed that only a straightforward conversion of dimensions has been used and no scaling factors. It is seen how the theoretical results obtained with 0.5-kg PV charges fit extremely well into the data obtained with charges orders of magnitudes greater.

The main result of the comparison between the standard row experiment with small charges and the large-scale experiment is that one recognizes the applicability of the theory, based on small-scale experiments, to the results from the large-scale experiments. The question of whether theoretical predictions may be used for extrapolation purposes seems thus to have been answered in the affirmative.

Having thus made the applicability of the theory plausible once more, the following remark may be made. Fig. 110b exhibits how the acceleration data are correlated with a broken line with different slopes in different intervals of the slant range. This empirically established behavior of the attenuation curve may now be replaced by the smooth theoretical curve. The physical reality behind it then becomes obvious in light of the discussion in Chapter 10.

THE SCALING PROBLEM

The desire to correlate experimental data from different experiments in different types of rocks and created by different types of explosives becomes understandable when for instance the results obtained in terms of the slope, λ, of the attenuation curves are considered. Table XXVIII shows the λ-values determined from experiments evaluated in the present text, compared with the result from the Cowboy

TABLE XXVIII

Comparison between experiments

Exp.	Rock	λ	κ	Remarks
F33		−1.7812		
F34	dolomite	−1.8250	0.022	
F33b		−1.7221		
F42	gneiss	−2.1646	0.075	Figs. 74 and 75
F47	granite	−1.5362	0.006	eq. 10.11
F50		−1.5185		eq. 10.12
F57	schist	−1.6816	0.004	eq. 11.4 } parallel to the lamination
		−1.5689		eq. 11.14 } parallel to the lamination
F67	gneiss	−2.2213	0.07	eq. 12.1
Cowboy	salt	−1.65		from Fig. 110a

experiment. As it is observed, the λ-values do not seem to vary too much (disregarding the data in gneiss) and, therefore, the idea of bringing the data together by multiplying them by properly chosen factors (scaling) may seem reasonable. The scaling factors which are proposed here are given in Fig. 110a for particle velocities and in Fig. 110b for particle accelerations. It is seen that the coefficients are 1.0 for salt and 1.3 for granite in both cases, but for tuff the coefficients are 5.0 for particle velocity and 14 for particle acceleration. No direct explanation for these factors is given and, what is more important, different coefficients must be introduced in the case of tuff for velocity and acceleration which seems to rule out any connection between these quantities. The evaluation of F57 in Chapter 11 showed how these should be interrelated, and it is now possible to scrutinize the idea of scaling in the light of the theoretical results.

First it is observed that the scaling procedure seems to be too simplistic an idea because it conceals the important information contained in the interdependence between the slope, λ, and the impulse, τ_0. This interdependence has been exhibited throughout the presentations in chapters 9, 10 and 11. The suggested composite attenuation curves have all the same slope regardless of the type of explosive used, a result that is contradicted by the results of F57. It also rules out the existence of types of rocks which exhibit a greater value $|\lambda|$ than the ones given. To bring this out more clearly attention is drawn to Fig. 111, where the result of F67 (confined charges) has been plotted into the same data as those shown in Fig. 110. Again it is seen that the results of small charges lie in the same region as the data from large-scale experiments, but now attention is fixed on the slope of the attenuation curve which in this case is very much different. This behavior can only be explained in the light of parameters κ and τ_0 as already shown.

Returning to the problem of the scaling factors, these may be interpreted by examining the results from the experiments discussed in this text with the following question in mind: What maximum pressure, p, is created by a detonating charge (e.g. 0.5 kg) at a given distance, d (e.g. 0.5 m), from its center? One may go back to each experiment and see what the answer to this question is. In Table XXIX the different experiments with 0.5-kg PV charges are listed together with one with 0.625 kg Am. The different types of rocks in which the experiments were performed are also given together with the λ-value obtained from the evaluations. By means of the equations already established, the maximum pressure, p (atm) at a distance of 0.5 m is computed and listed for each case. It is seen that p varies greatly from case to case. It is also noticed that no apparent interdependence between the type of rock and the value of p created in it can be detected. The variation in p expresses a real physical difference between the cases. It is mainly caused by the variation in the rock's resistance to being crushed and pulverized in the immediate neighborhood of the explosion and in the action of the explosive. To some extent this is accounted for in the way the distances are non-dimensionalized, but it is unavoidable that local differences in the rock's properties will affect p beyond that. If namely the introduction of the "slant range" accounted for the whole difference, there would be no need for additional "scaling coefficients" which are dependent on the rock type. The cases F42 and F67, in which the rock was gneiss, show λ-values which are almost identical, but the p-values differ by a factor of 2. These considerations now explain the background for the scaling coefficients introduced in Fig. 110a and b. It also explains, in view of the theoretical considerations, why different scaling "coefficients" may have to be used for particle velocity and particle acceleration. This insight can, however, only be obtained through theoretical considerations, whereby the physical reality behind them is brought to light which may have been obscured by a straightforward reliance on "composite" curves. Further comments on this point will be found in the next section.

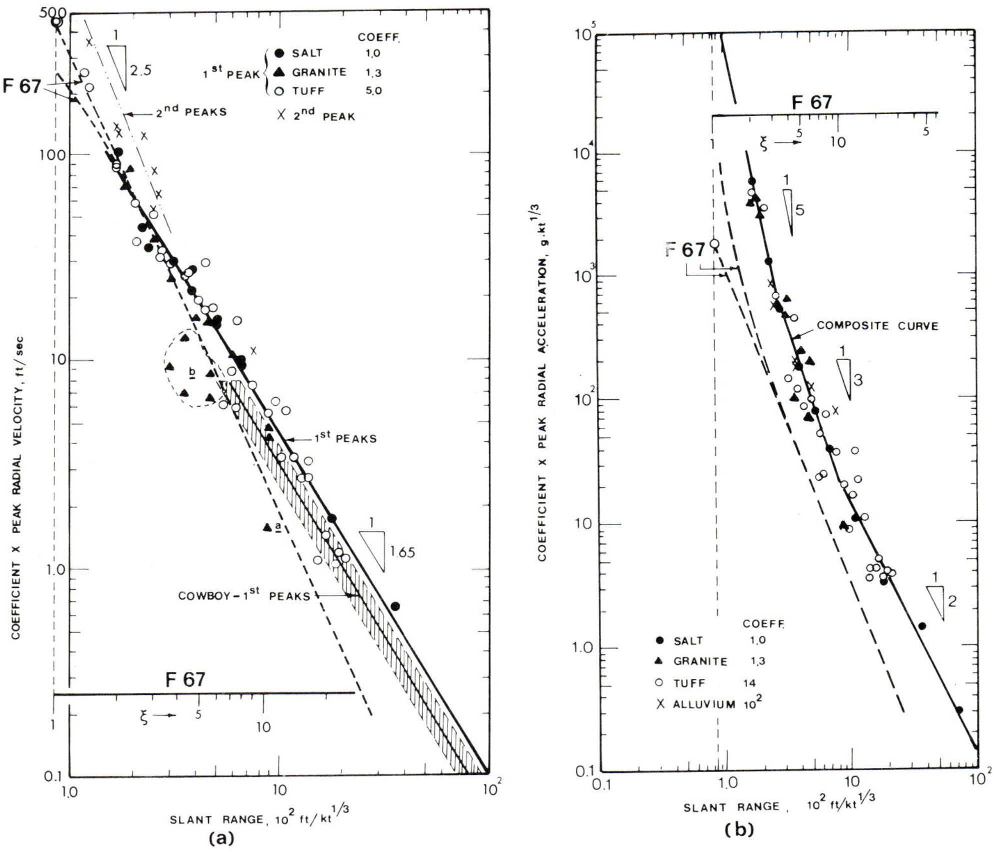

Fig. 111. Replot of data from large-scale experiments with the results of F67 drawn for comparison (confined charges).

TABLE XXIX

Comparison of experiments

Exp.	Rock	λ	p(atm)	Remarks
F33	dolomite	−1.7812	615	
F34		−1.8250	574	
F42	gneiss	−2.1646	764	
F47	granite	−1.5362	472	
F50		−1.5185	286	
F57	schist	−1.6816	193	0.5 kg PV
		−1.5689	(144)	0.625 kg Am
F67	gneiss	−2.2213	393	

THE EXTRAPOLATION PROBLEM, 2 (SAFE DISTANCES)

The extrapolation problem was treated in part 1 (p. 195) and the main purpose then was to ascertain that small-scale experiments produced relevant data for large-scale explosions. Now when this has been made plausible, the practical implications of this fact must be brought out. As mentioned in the previous section this touches also on the usefulness of the so called "composite" attenuation curve. The implications of the theoretical approach are best shown by an example.

Assume that a confined charge of 100 kg PV is to be detonated in the neighborhood of an underground facility. What is then the smallest distance from this facility at which this explosion may take place without endangering it? This is the classical question of the "safe distance", and in the following certain aspects of the determination of safe distances will be discussed.

First it is noticed that the question of the sensitivity of the facility will have to be dealt with. Sensitivity criteria of underground structures may be of different types. A survey of some of the considerations which are relevant in this connection is given in Chapter 14. For the present purpose it is assumed that the facility can sustain the influence of an oncoming stress wave if its maximum amplitude $A \leqslant 20$ atm. The sensitivity criterion is in other words given here in terms of the peak pressure of the stress wave.

The second piece of information necessary for the determination of the safe distance is the characteristics of the rock in which the facility is located. In this example it will be assumed that the facility is located in the rock tested in F57. One has thus what may be considered a complete knowledge of the rock's characteristics, and one may use this knowledge to extrapolate from the 0.5-kg PV charges to the 100-kg PV charges.

From 11.18 one obtains κ and τ_0:

$$\kappa = 0.004, \quad \tau_0 = 0.3 \tag{13.1}$$

From Table XVII one then obtains

$$\lambda = -1.67322, \quad B = 1.87532 \tag{13.2}$$

and from 11.21 the value of the characteristic length a is obtained:

$$a = 0.348 \text{ m} \quad (0.5 \text{ kg PV}) \tag{13.3}$$

This is sufficient information for plotting the non-dimensional stress, S_{max}, as a function of the non-dimensional distance, ξ, as done in Fig. 112. Now, all data given in 13.1, 13.2 and 13.3 are valid for 0.5 kg PV. Presumably, the non-dimensional quantities κ, λ and B will be the same also for 100 kg PV. Furthermore, τ_0

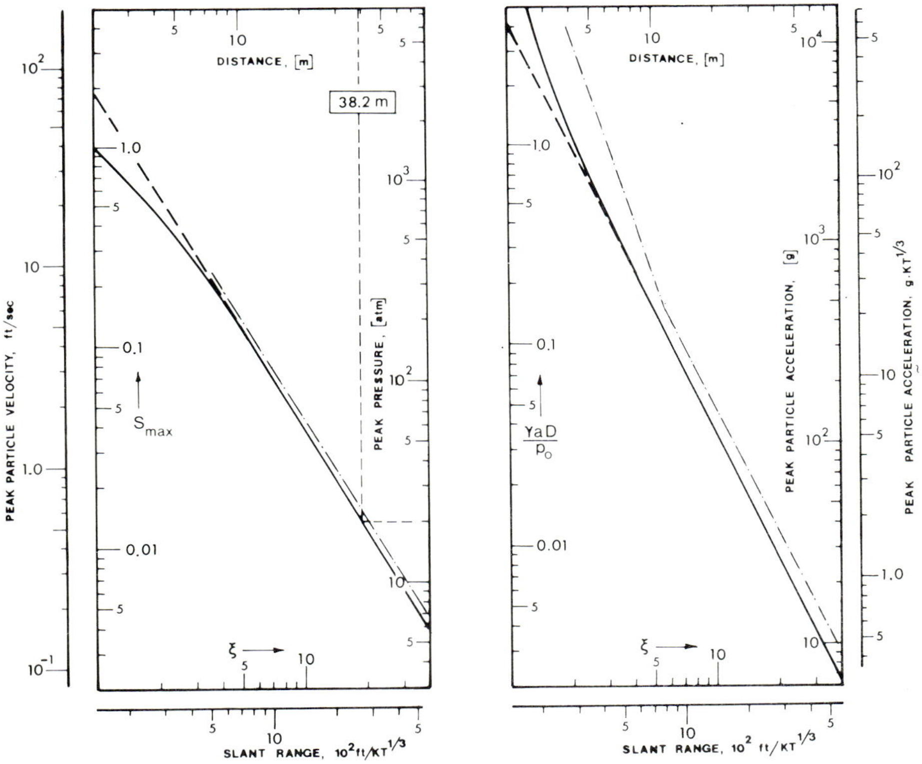

Fig. 112. Determination of the "safe distance" through application of the attenuation curves (100 kg PV). Solid curve: theoretical curve; dashed curve: straight-line approximation; dash-dotted curve: "composite" curve.

will remain the same assuming the validity of the conclusion drawn on p. 140. The only quantity which then is changed is the characteristic length a, which for 100 kg PV may be computed on the assumed validity of 10.3:

$$a = 0.348 \left(\frac{100}{0.5}\right)^{1/3} = 2.035 \text{ m} \qquad (100 \text{ kg PV}) \qquad [13.4]$$

This then is sufficient information for the determination of the dimensional distance as shown by the distance scale in Fig. 112.

The peak pressure at $\xi = 1$ is p_0, which is obtained from Table XVIII. It must, however, be remembered that this value is based on the value $a = 0.4$ m for the characteristic length. Thus:

$$p_0 = \left(\frac{0.348}{0.4}\right)^{\lambda} \cdot 1181.13 = 1491 \text{ atm} \qquad [13.5]$$

With this value of p_0 the peak pressure scale of Fig. 112 is drawn.

For comparison the composite curve of Fig. 111 is plotted (dash-dotted curve) and Fig. 112 also shows the particle acceleration curves. The details of this diagram are left for the reader to contemplate.

The safe distance can now without difficulty be extracted from Fig. 112 and, as shown, this distance is 38.2 m. It is left to the reader to find out from the diagram the peak particle acceleration which the facility will experience.

One might now extend the scope of the present example by asking: What difference would it have made if the charge had been 100 kg Am instead of PV? Because the experiment F57 also was conducted with this type of explosive, the answer can readily be given.

From 11.18:

$$\kappa = 0.004, \quad \tau_0 = 0.5 \qquad (0.625 \text{ kg Am}) \qquad [13.6]$$

From Table XVII:

$$\lambda = -1.57661, \quad B = 2.14107 \qquad (0.625 \text{ kg Am}) \qquad [13.7]$$

From 11.21:

$$a = 0.3845 \text{ m} \qquad (0.625 \text{ kg Am}) \qquad [13.8]$$

Assuming again the validity of 10.3 one obtains:

$$a = 0.3845 \left(\frac{100}{0.625}\right)^{1/3} = 2.087 \text{ m} \qquad (100 \text{ kg Am}) \qquad [13.9]$$

Finally p_0 is found from Table XVIII where the value refers to a characteristic length $a = 0.4$ m, i.e.:

$$p_0 = \left(\frac{0.3845}{0.4}\right)^{\lambda} \cdot 715.53 = 761.53 \text{ atm} \qquad [13.10]$$

This gives all the necessary information for plotting the attenuation curves in Fig. 113 from which the safe distance now is determined as 33.5 m.

The final change in the present example arises from the question: What difference would it have made if the rock had had other characteristics? This can be answered by assuming that the facility under investigation was located in the rock tested in experiment F67. The procedure is the same as before.

From 12.4:

$$\kappa = 0.072, \quad \tau_0 = 1.57 \qquad (0.5 \text{ kg PV}) \qquad [13.11]$$

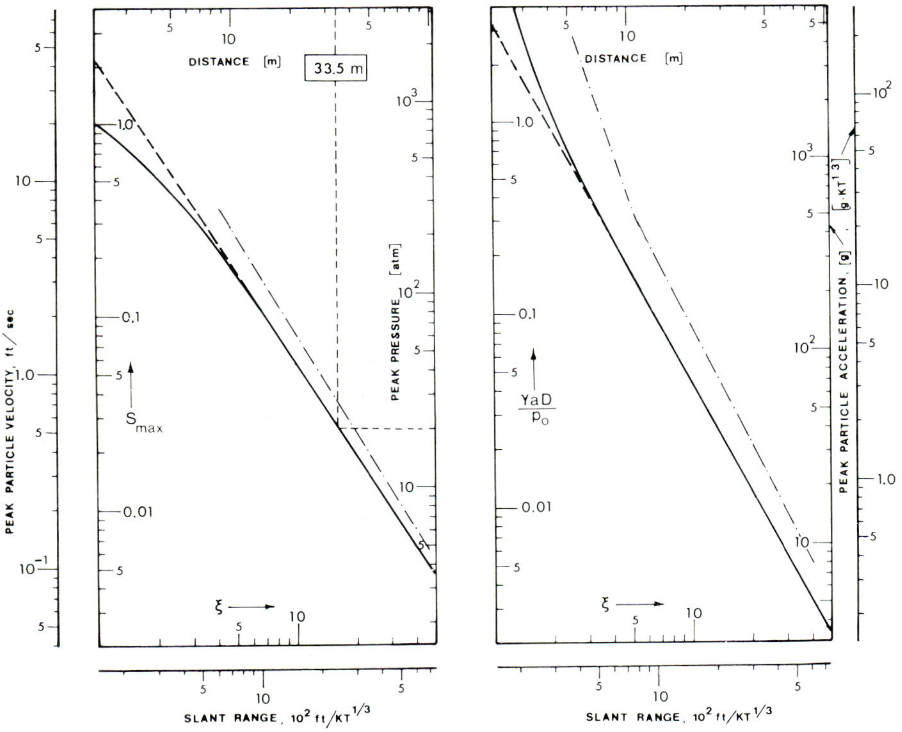

Fig. 113. Determination of the "safe distance" through application of the attenuation curves (100 kg Am). Solid curve: theoretical curve; dashed curve: straight-line approximation; dash-dotted curve: "composite" curve.

From Table XXVII:

$$\lambda = -2.21890, \quad B = 0.79263 \qquad (0.5 \text{ kg PV}) \qquad [13.12]$$

From 12.14:

$$a = 0.2042 \text{ m} \qquad (0.5 \text{ kg PV}) \qquad [13.13]$$

Assuming again the validity of 10.3 one obtains:

$$a = 0.2042 \left(\frac{100}{0.5}\right)^{1/3} = 1.194 \qquad (100 \text{ kg PV}) \qquad [13.14]$$

Finally p_0 is found from 12.15:

$$p_0 = 15014/B = 8375 \text{ atm} \qquad [13.15]$$

This completes the necessary information for the determination of the attenuation curve in Fig. 114 from which the safe distance in this case is found to equal 22.7 m.

To conclude this example we consider the many different aspects of the extrapolation problem in the light of what has so far been achieved.

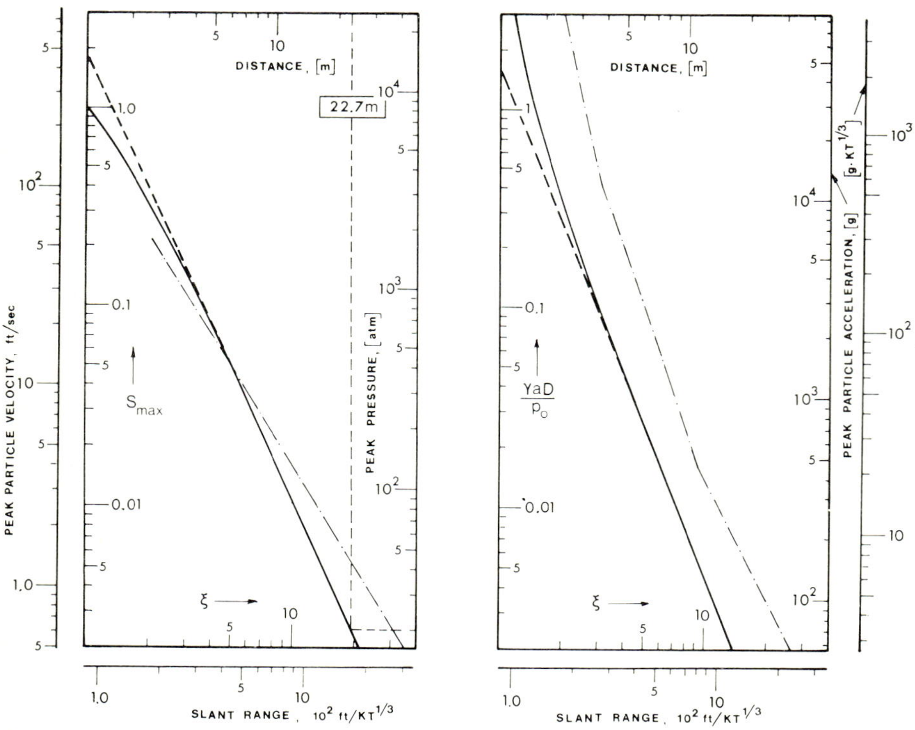

Fig. 114. Determination of the "safe distance" through application of the attenuation curves (100 kg PV). Solid curve: theoretical curve; dashed curve: straight-line approximation; dash-dotted curve: "composite" curve.

(1) It has been demonstrated that the process which takes place when a confined charge is detonated and stress waves are created in a specific type of rock is characterized by parameter κ. This parameter thus plays a similar role as a Reynolds number in fluid mechanics.

(2) Furthermore, it has been demonstrated that parameter τ_0 exhibits the influence of the type of explosive used. However, experimental evidence indicates that the so-called TNT-equivalence of an explosive will depend on the rock, and that it may actually change from one rock site to another even though the rock, from a geological point of view, remains the same. Parameter τ_0 takes this effect into account, and exerts its influence on the slope λ of the attenuation curves.

(3) In addition, the two quantities a and p_0, i.e. the radius of the imaginary sphere and the pressure acting inside it, become very important in the extrapolation procedure as has just been demonstrated by the examples in Figs. 112, 113 and 114. The fact that p_0 may differ from one type of rock to another, even though all other parameters may remain unchanged, explains in view of the normalizing factors used in the attenuation curves for particle acceleration why different coefficients may have to be introduced for particle acceleration and particle velocity, as is the case for tuff in Fig. 110.

Now, all these parameters, a, p_0, κ and τ_0, are determined by a standard row experiment. It thus becomes evident, that a standard row experiment may serve as a "material testing experiment" upon which predictions may be based. In this way the engineering predictions in this field are brought on the same basic foundation as engineering predictions in other fields. It is also clear that the idea of composite attenuation curves is too simplistic because it tends to obscure the physical realities pointed out here.

Chapter 14

SENSITIVITY CRITERIA FOR TUNNELS

INTRODUCTION

In the applications of the theory discussed in the preceding chapter, the question of the sensitivity of an installation in a rock site arose. With more and more underground facilities being built and the coming demand for expansion of such facilities, the question of how to enlarge an underground facility without destroying it becomes increasingly important. A proper understanding of the interaction between tunnels and shock waves is imperative in this connection.

It is realized, that the problems touched upon here cover a very large variety of possible situations, and even though some general features may remain the same, each situation seems to require separate handling. Because of the variety of constellations one may have to consider, the present approach can only be considered as an attempt to discuss a few selected problems.

A cavity's or a tunnel's capacity to withstand the influence of an oncoming shock wave without suffering serious damage depends on a number of factors; some of them are not connected with the shock wave itself. There is a great difference between a lined and an unlined tunnel. For an unlined tunnel the method used for its excavation determines the extent to which the neighboring zone of rock has been dislocated, and this in turn has a great influence on the sensitivity of the tunnel to the influence of a shock wave. Another example of an influencing factor is the static stresses around the tunnel. One must realize, that these may be so great, that the tunnel may collapse or suffer serious damage from relatively weak shock waves. Thus, the collapse criterion for a tunnel, whether lined or unlined, will have to be considered in the light of information unobtainable from the shock wave alone.

Now, given the situation of an extension of an already existing underground facility, it may be that the tunnel is equipped with instruments or other installations which may suffer damage long before the tunnel itself is endangered. Thus, the sensitivity criterion for the facility is determined from considerations which again have no bearing on the rock itself. However, any influence on installations in the tunnel will have to be transmitted to the installations through the tunnel walls, and it becomes imperative to be able to predict the behavior of the walls when hit by an oncoming shock wave in cases when the walls do not suffer damage.

These comments show the necessity of limiting the exposition in this introductory text to certain well-defined problems. The attention will here be focused on lined tunnels only, and the results to be shown will assume the absence of static stresses. These will eventually have to be superimposed on the dynamically created stresses. Two cases will be discussed, namely the collapse of the tunnel and the wall velocities of the tunnel when it is left unharmed by the shock wave.

DEFINING THE PROBLEM

The interaction between a cylindrical cavity in a given material and an oncoming stress wave transmitted through the same material has been the topic of several theoretical investigations. Here only two contributions (Baron and Matthews, 1961; Baron and Parnes, 1962) are mentioned as typical examples of the approach which is applied. Even though these give some answers and thus furnish some insight into the physical realities of the interaction, key questions, such as the influence of different types of lining, different types of rock, etc., remain unanswered. Such analytical investigations may be complemented by modern numerical techniques using electronic computers, but even then the input must be made by means of experimentally obtained information. The present approach therefore will exclusively be based on experimental evidence.

The first thing to realize is that a tunnel's lining may be constructed in many ways, from linings which are only layers of concrete which have been sprayed on to the raw tunnel walls to the most sophisticated designs of spring-supported inner structures. The most commonly encountered linings are, however, the ones exhibited in Fig. 115. The lining in Fig. 115a is to be considered as a cylindrical shell of reinforced concrete, where presumably a good connection between the rock and the concrete will have to be ensured through practical means during construction. Fig. 115b shows the same type of lining where anchors have been added to increase

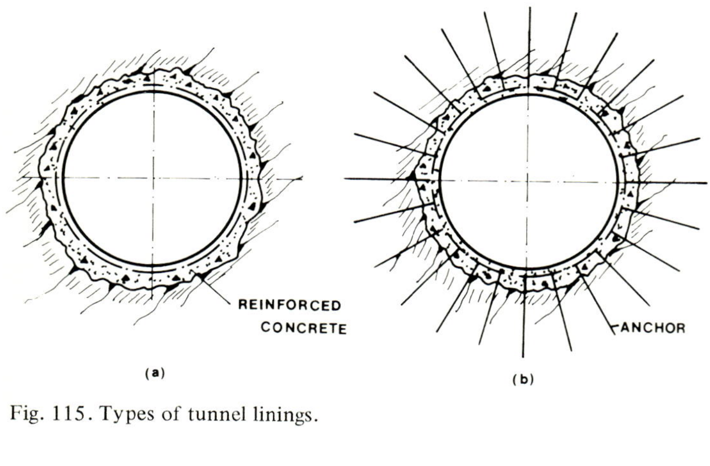

Fig. 115. Types of tunnel linings.

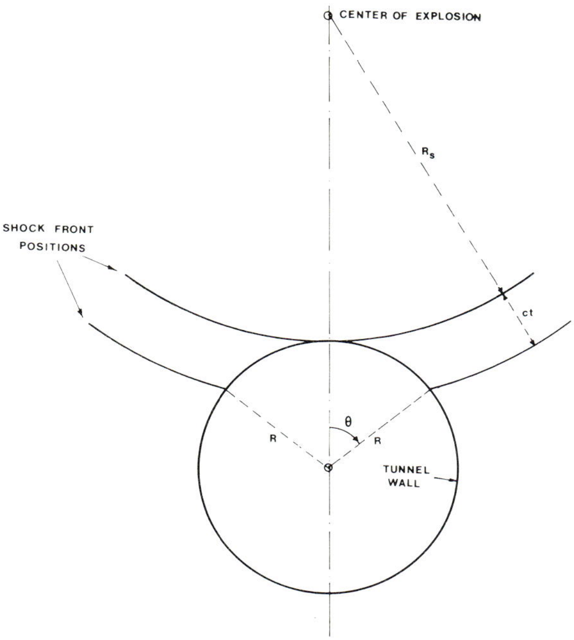

Fig. 116. Stress wave hitting a cylindrical cavity.

the strength of the lining. One should not, however, anticipate that the anchors will have too much influence on the way in which the cavity behaves when struck by a stress wave. Even though such anchors may be pre-stressed, their main value seems to lie in their preventing a collapse to be totally disasterous, i.e. a complete cave-in.

Having thus limited the present discussion to a certain type of tunnel lining, one may proceed to review the problem as it appears in practice. The oncoming stress wave is considered to be created by a spherically symmetric explosion, i.e. it will be a spherical wave as exhibited in Fig. 116. As the stress wave moves away from the center of explosion, it has its maximum amplitude attenuated and exhibits a dispersion prior to the time when the shock front reaches the tunnel. As it envelopes the cavity, a rather complex reflection procedure takes place and surface waves along the wall will be created. The simplified approach which analytical and/or numerical techniques must one way or another have built into them, will in no way be able to account for the complexity of reality. Even experimental evidence from real tests will turn out to be inadequate in this respect. In the following,the scope of experimental investigations will therefore be narrowed down to those main problems which may carry the necessary information for obtaining the sensitivity criteria and which also are believed to have general applicability.

A third feature of the problem is now brought to the foreground. A dimensional

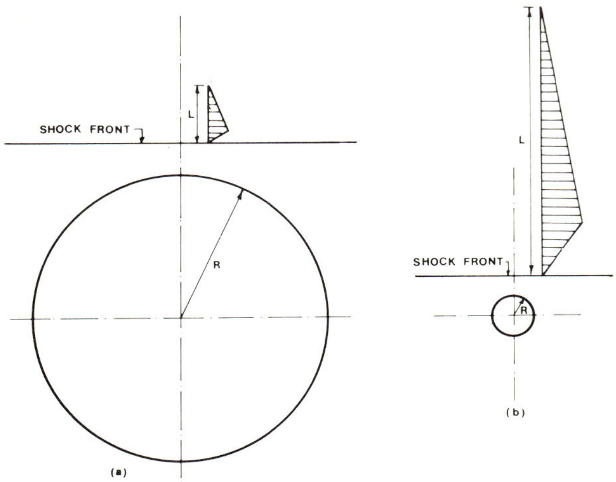

Fig. 117. Comparison between the shock wavelength, L, and the cavity radius, R, for two extreme cases.

analysis of the problem will show that the ratio between the radius, R, of the tunnel and the wavelength, L, of the shock wave will appear as a parameter of the problem. Fig. 117a illustrates the situation for which $R/L \gg 1$, a case which will be totally different from the case in Fig. 117b where $R/L \ll 1$. In the latter case the stress distribution around the cavity is often described by the stress concentration factors for a circular hole in a plane stress field, an elementary solution which is supposed to be applicable in an assumed semi-steady situation. Both cases in Fig. 117 are, however, extremes because the situation where R and L are of the same order of magnitude is the one most often relevant in practical cases. Thus, the problem has been narrowed down to the investigation of tunnels with one specified type of lining hit by a stress wave with a wavelenght of the same order of magnitude as the radius of the tunnel, i.e. $1/10 < R/L < 1$.

The limitations on the scope of the present text set down in these preliminary remarks makes it necessary to elaborate somewhat on the consequences. It is noted that the stress waves exhibited in Fig. 117 are only shown with their first compression phase. The following tension phase is here neglected, the assumption being that this first compression phase is the one that will endanger the tunnel. Under certain circumstances the following tension phase may also have to be considered, but this case is not treated here due to the lack of experimental evidence.

THE COLLAPSE OF A TUNNEL

The collapse of a tunnel due to an oncoming stress wave will here be studied in the light of the results from an experiment. The layout was such that in all seven tunnels with different types of lining were hit by one and the same shock wave. The tunnels were of different size, and the linings used were the ones shown in Fig. 115. In addition, tunnels with linings of concrete with no reinforcement and tunnels with no linings were also tested. The charges used were great enough to bring all

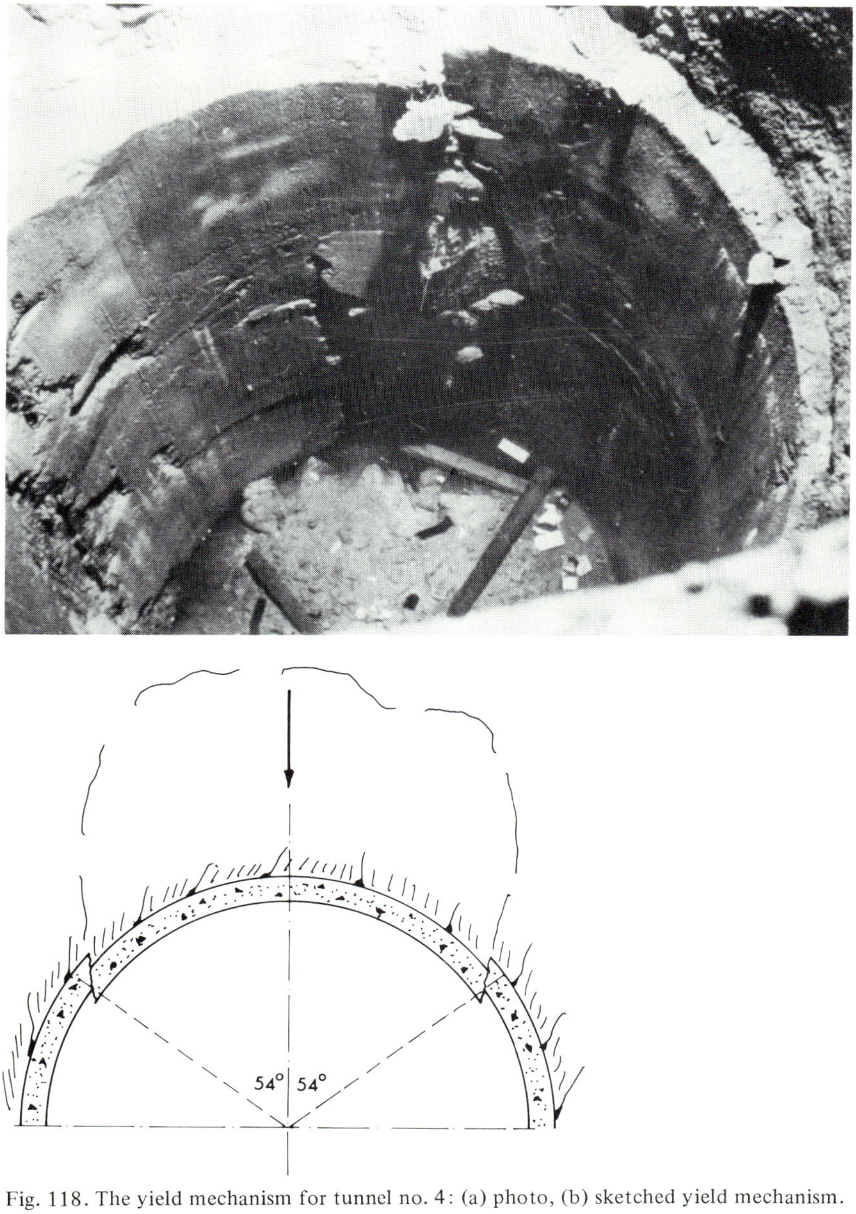

Fig. 118. The yield mechanism for tunnel no. 4: (a) photo, (b) sketched yield mechanism.

lined tunnels to collapse. The tunnels with no linings survived without damage because they were far enough away from the charge so that the stress wave had been attenuated below the danger point of the tunnels. The tunnels with linings according to Fig. 115 suffered damage such as shown in Fig. 118 for tunnel no. 4. This tunnel had no anchors, and the lining yielded along generatrices of the cylindrical tunnel for which $\theta \simeq \pm 54°$, θ being given as shown in Fig. 116. The yield is apparently caused by shear and the lining and a part of the surrounding rock are displaced inwards as shown. The yield zones are shown in Fig. 119 as a mount of

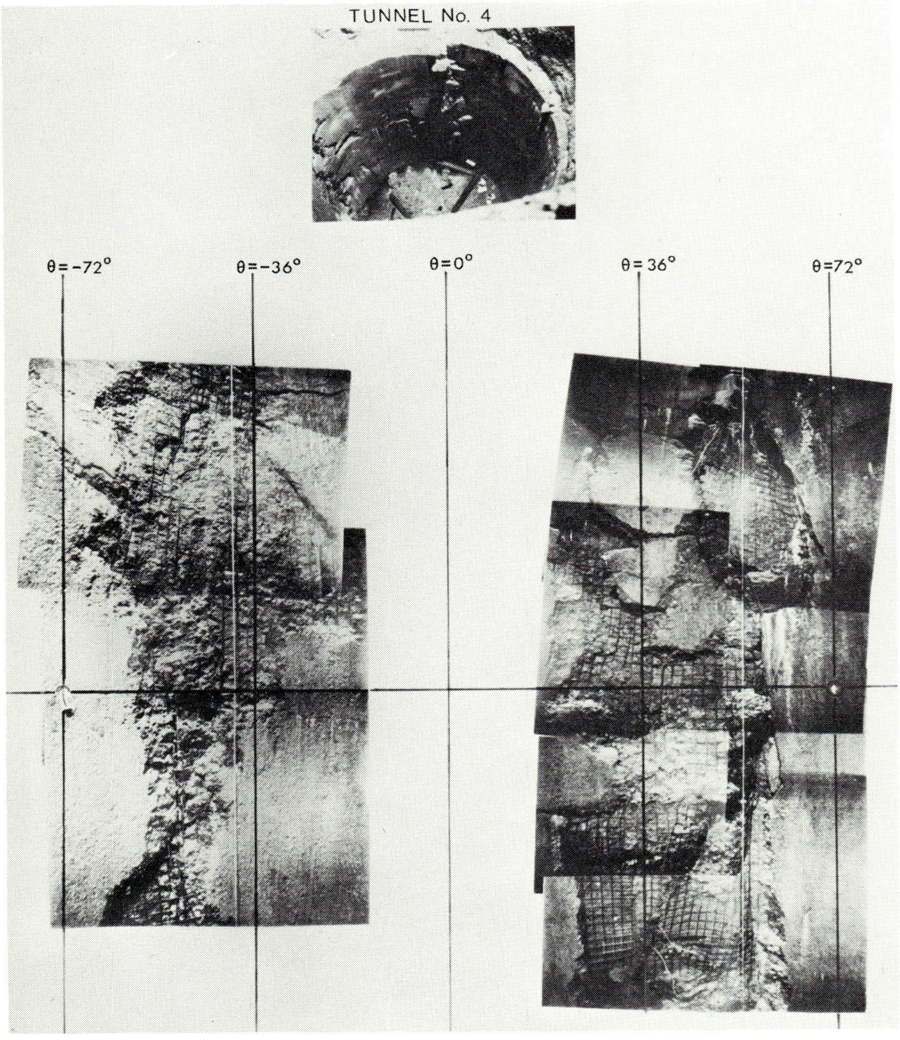

Fig. 119. Mount of single photos showing details of the yield zones (tunnel no. 4.)

single photos pieced together. The lines drawn in are the generatrices of the tunnel indicating the positions of the accelerometers which were fixed to the wall to determine the wall velocities at these positions in the radial direction. The S-shaped configuration of the reinforced bars after yield is shown in Fig. 120.

Tunnel no. 5 was lined identically to tunnel no. 4, the only difference being the addition of anchors as shown in Fig. 115b. The yield occurred in the same way as described for tunnel no. 4 but the value of θ at which yield occurred was now estimated at $\theta = 72–75°$. The yield surfaces in this case are shown as a mount of single photos in Fig. 121.

The observations of the yield serve now as a starting basis for contemplating the yield mechanism. First it should be observed, that the stress waves used to cause collapse of the tunnels had a wavelength L which gave the ratio $R/L \cong 1/10$. Under such circumstances the collapse of the tunnel occurs during the first impact of the wave on the tunnel, i.e. one concludes that the first compression phase of the shock

Fig. 120. S-shaped reinforcement bars after yield (tunnel no. 5).

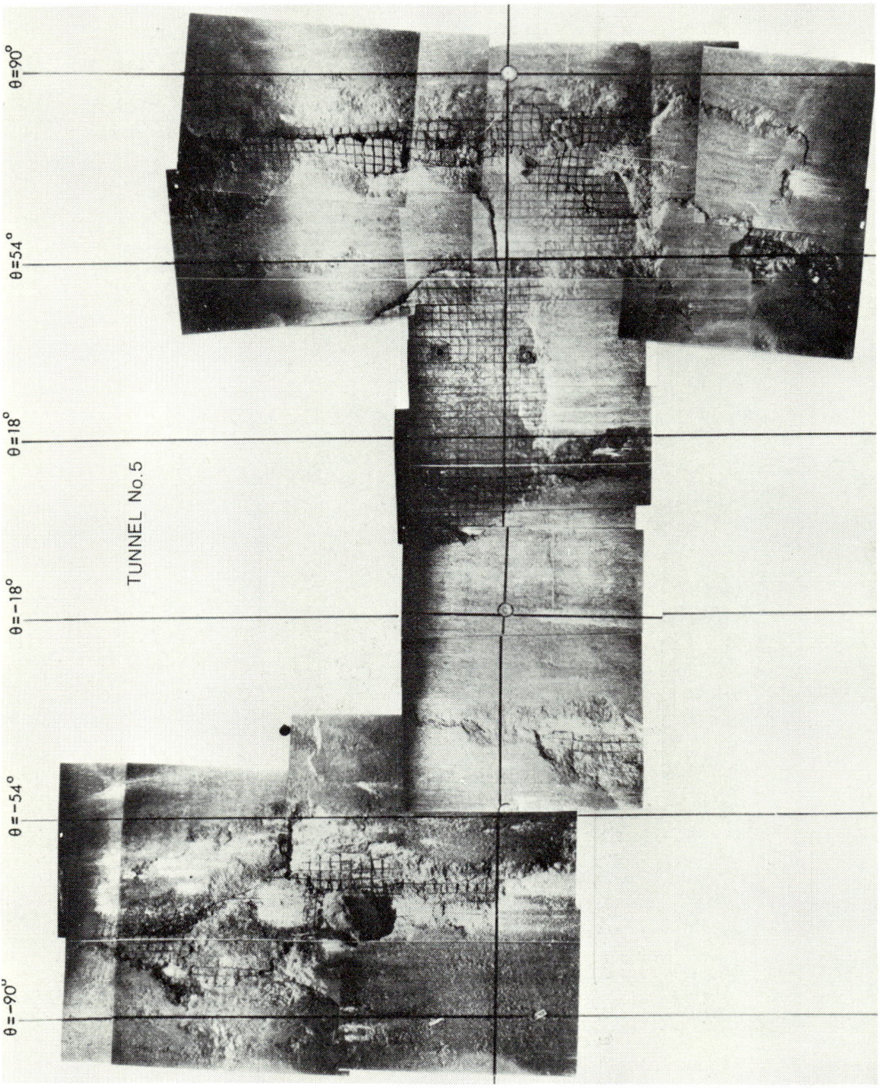

Fig. 121. Mount of single photos showing details of the yield zones (tunnel no. 5.)

wave represents the endangering loading on the tunnel. The next step is to conclude, that since the lined tunnel apparently can withstand the influence of a shock wave, which an unlined one does not, then the increased capacity of the lined tunnel can only be attributed to the carrying capacity of the lining. This lining can then either be considered as a shell or as an arch. Fig. 122 shows a sketch of the lining at the moment that the tunnel is engulfed by the shock wave. It will act as a circular arch over a steadily increasing angle $\theta = \theta_0$. The calculation of stresses in this arch is hampered by the fact that, to some extent, it is prevented from deforming outwards, and also the exact loading, which is here time dependent, is not known.

An exact analysis of the problem is beyond the scope of this presentation, and in the following two simplifying assumptions will be made, which to some extent give a plausible explanation of the collapse mechanism. It will be assumed that the

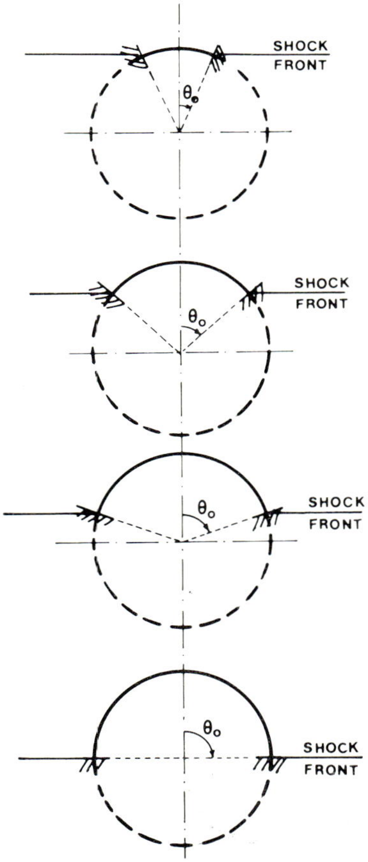

Fig. 122. Sketch showing the lining acting as a clamped circular arch with increasing angle (θ_0) as the shock front progresses.

elementary calculation of stresses in an elastic arch is applicable. Secondly it will be assumed that a quasi-static approach may be used, i.e. the stresses are computed for

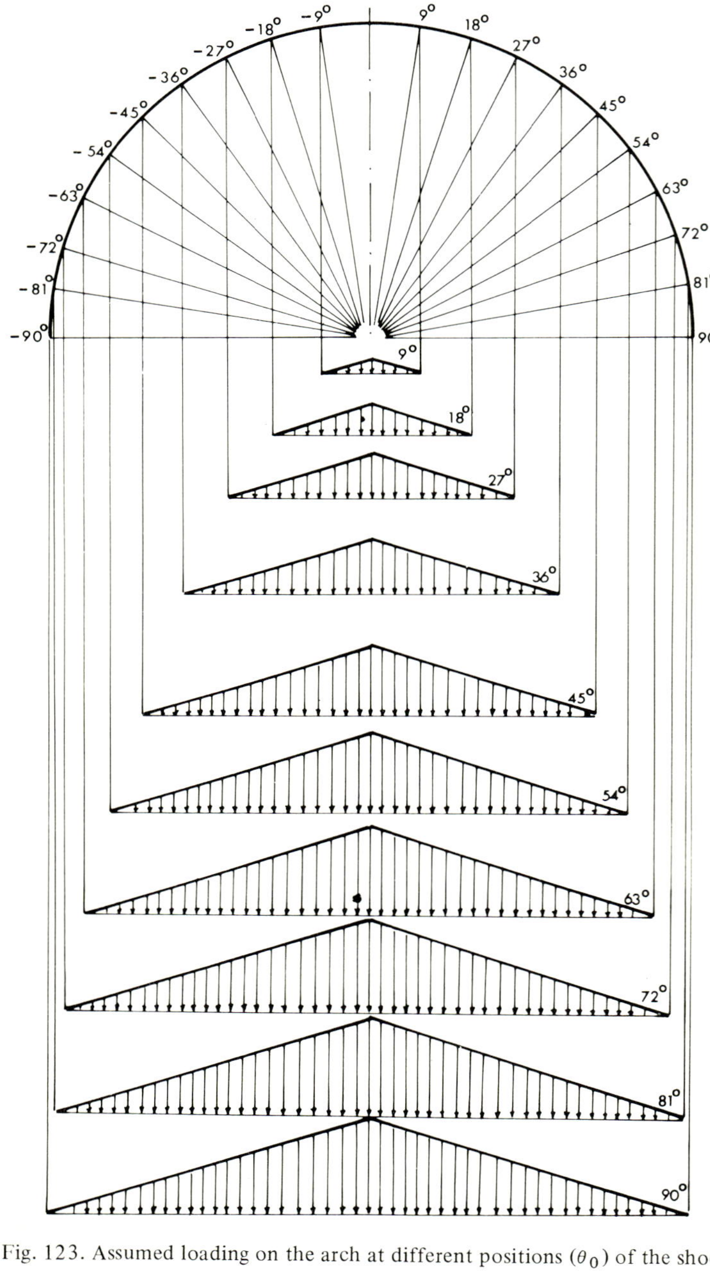

Fig. 123. Assumed loading on the arch at different positions (θ_0) of the shock wave.

given positions (θ_0) of the shock wave assuming at each position a static load acting on the arch as indicated in Fig. 123. An analysis based on these assumptions, with the following load function, p:

$$p = p_0 \, (\sin \theta_0 - \sin \theta) \qquad [14.1]$$

may then easily be carried out.

The result of such an analysis is displayed in Figs. 124 and 125. Fig. 124 shows the direction of the principal axes 1 and 2 at the centerline of the clamped-in section of the arch, i.e. at the position θ_0 of the shock wave. The arrows show the direction of the shear force, Q, at a 45° angle with these axes. This shear force has its maximum value at $\theta_0 = 71°$. However, a closer analysis shows that the absolute greatest shear force at a 45° angle with the principal axes occurs at $\theta = 64.8°$ when the position of the shock wave is given by $\theta_0 = 81°$. This is displayed in Fig. 125 which also shows with bold lines the regions at the outer fibers of the arch where tensile stresses will occur. These regions will, however, depend on the ratio R/h where h is the thickness of the lining.

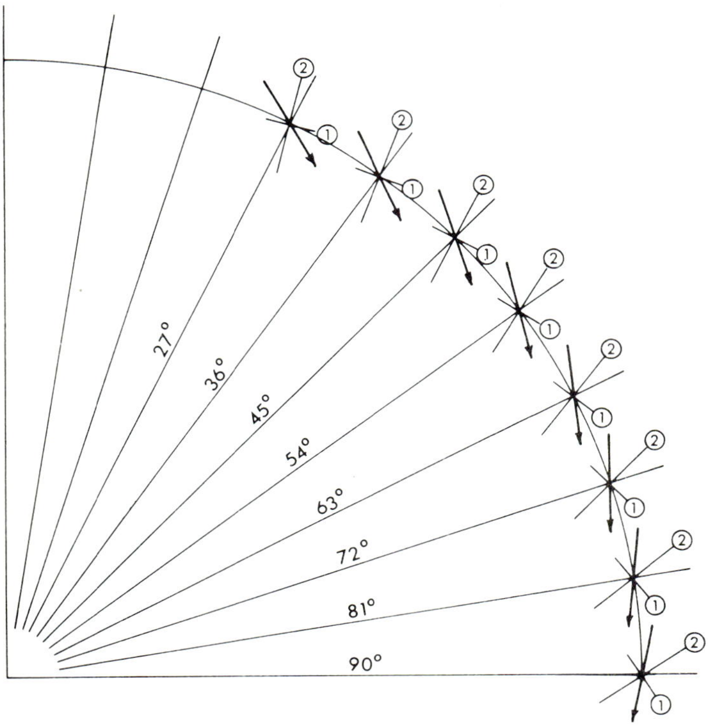

Fig. 124. Principal axes 1 and 2 at the center line of the arch computed at $\theta = \theta_0$ for different values of θ_0.

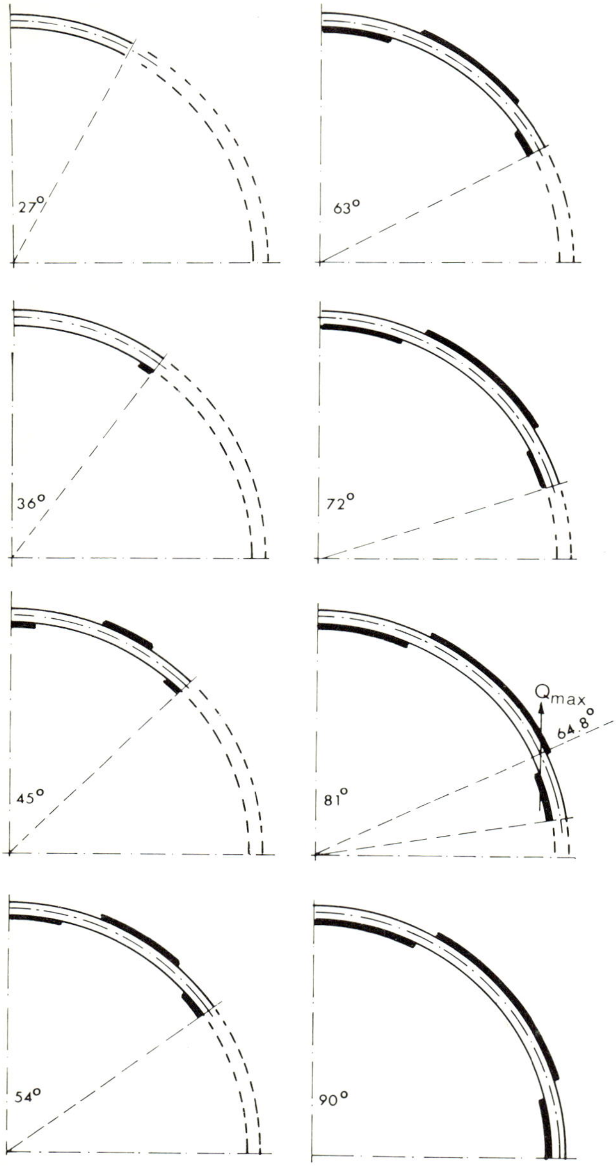

Fig. 125. Regions of tensile stresses at the outer fibers of the arch shown with bold lines for different positions (θ_0) of the shock wave.

If these results are confronted with the experimental evidence, the conclusion which may be drawn is the following. The collapse mechanism may be conceived of

Fig. 126. Yield of a concrete prism.

as an ordinary yield of a concrete prism as shown in Fig. 126. The yield will take place mostly in the region where the maximum shear force occurs, i.e. according to the present analysis in the region given by $60° < \theta < 81°$. This corresponds very well with the observed yields in the cases mentioned. Thus one has here been able to demonstrate, without any claim on accuracy being made, the type of considerations involved in establishing the collapse criterion for a tunnel.

WALL VELOCITIES

In cases when the sensitivity of an underground facility depends on the different types of installations in it, the knowledge of the motion of the tunnel walls during passage of the stress wave becomes important. Again the scope of this presentation is narrowed down to a specific case which exhibits to some extent the considerations involved.

Returning for a moment to Fig. 116, it is realized that the maximum amplitude of the stress wave will be attenuated when travelling over a distance equal to the diameter ($2R$) of the tunnel. This attenuation will depend on the explosion which created the stress wave, because it is expressed in terms of the non-dimensional distance ξ. In this way the characteristic length a of the explosion must be accounted for when discussing the oncoming stress wave.

For cases within the range of R/L considered here, the wall velocity for values of θ in the range $-180° < \theta < -90°$ and $90° < \theta < 180°$ are negligible. It means that the "back" side of the tunnel is for all practical purposes unaffected by the stress wave, it lies in the "shadow zone". (It should be noted that this statement is based on the assumptions made, and that cases may occur where the shadow zone is indeed affected.) The oncoming stress wave will be reflected from the tunnel walls. The details of this rather complex reflection process is here avoided by using experimentally obtained data to connect the measured wall velocities with the characteristic data of the oncoming stress wave. These are assumed known from the theoretical description.

The oncoming stress wave is plotted in Fig. 117 and the assumed triangular form of its first phase is known if its length L, its maximum amplitude A and its dispersion (tg α) are known. One obtains, however, a more adequate picture of the oncoming stress wave if its effect is plotted as particle velocity v_r versus time t as shown in Fig. 127. Its duration is described by L/c and its maximum amplitude A is given as particle velocity (m/s). This then is regarded as the input from which the wall velocity must be deducible.

The wall velocity is here to be described as a function of time t and position θ.

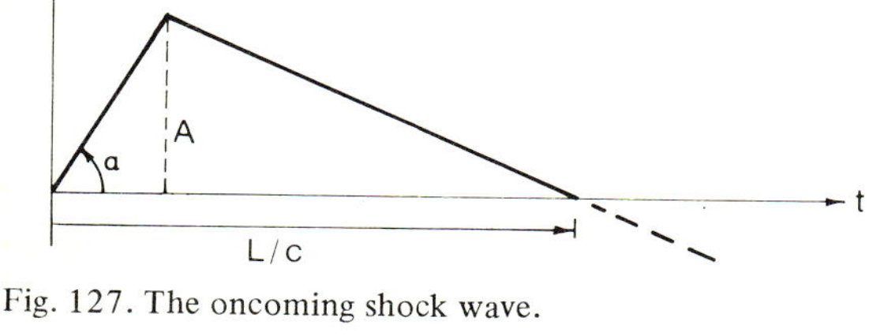

Fig. 127. The oncoming shock wave.

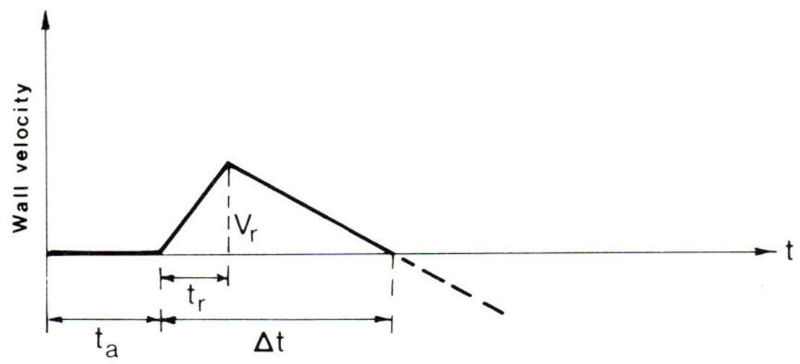

Fig. 128. Radial wall velocity as a function of time.

The fact that the results presented here are empirical formulae deduced from experimental evidence puts a limitation on their applicability. The first limitation lies in the already stated range of the ratio R/L. The second will be the assumption that the main influence on the tunnel walls will be experienced during the impact of the first phase of the stress wave on the tunnel. From this follows that only the first phase of the wall velocity will be considered here, a limitation which is also set by the reliability of the experimental evidence on which the empirical formulae are built. The velocity in the radial direction of the wall at a given position θ is shown in Fig. 128. Its shape is assumed to be triangular with sufficient accuracy. The time $t = 0$ denotes the time when the front of the shock wave hits the tunnel at $\theta = 0$ (Fig. 116). The arrival time t_a of the shock wave at position θ, the duration, Δt, of the positive phase, the maximum value, V_r, of the positive phase and the rising time, t_r, are now the four quantities which determine the wall velocity completely. These four quantities are specified below.

From Fig. 116 one obtains immediately from geometrical considerations:

$$\frac{ct_a}{R} = \frac{R_s}{R}\sqrt{1 + 2\left(\frac{R}{R_s} + \frac{R^2}{R_s^2}\right)(1 - \cos\theta)} - 1 \qquad [14.2]$$

where c as before is the signal velocity of the rock. When the radius of curvature, R_s, of the shock is large compared with the radius, R, of the tunnel, the simplified formulae (for a plane shock) may be applied:

$$\frac{ct_a}{R} = 1 - \cos\theta \qquad [14.3]$$

It should, however, be mentioned, that this purely geometrical consideration with an assumed known value of c from free field experiments must be somewhat adjusted due to the reflection of the shock wave from the tunnel walls. It seems that a reduction of c by approximately 25% renders the best agreement with observations.

The peak wall velocity, V_r, has a θ-dependence which may empirically be expressed by:

$$V_r = 2A \exp\left[-\left(\alpha \frac{\theta}{180^\circ}\right)^2\right] \qquad [14.4]$$

where θ is to be introduced in degrees, and α is a constant. It is noticed that $V_r = 2A$ for $\theta = 0$ which corresponds to the known reflection process for particle velocities discussed in Chapter II, Fig. 11. The constant α accounts for the previously discussed influence of the free field attenuation of the shock wave as it travels a distance corresponding to the positions $\theta_0 = 0$ to $\theta_0 = 90^\circ$. This influence is accounted for as follows: Let S_0 and S_{90} be the non-dimensional peak stresses of the shock wave at the positions corresponding to the values of θ_0 above. Then:

$$\alpha = 1.784\, S_0/S_{90} + 0.256 \qquad [14.5]$$

It is clear that 14.4 gives V_r as a "hat curve" with a width that varies with the way in which the shock wave is created.

The duration, Δt, of the positive phase of the wall velocity is given by:

$$\Delta t = \Delta t_0 \exp\left[-\left(\beta \frac{\theta}{180^\circ}\right)^2\right] \qquad [14.6]$$

where β is a constant and where Δt_0 is connected to the input data of the stress wave by:

$$\Delta t_0 = \frac{L}{c} \tanh\left(\frac{5R}{L}\right) \qquad [14.7]$$

The constant β is thought to be uninfluenced by the stress wave and is:

$$\beta = 2.25 \qquad [14.8]$$

The rising time, t_r, of the positive phase of the wall velocity is given as:

$$\frac{t_r}{\Delta t} = 0.059 + 1.510 \left(\frac{\theta}{180^\circ}\right)^2 \qquad [14.9]$$

One has now all the necessary information for a complete description of what happens at the wall during the first positive phase. It is stressed, that these formulae are empirical, and that they are deduced from experimental results which limits their applicability to the range covered by the experiments. It is, however, thought that they give a fair starting point for the considerations involved in establishing the sensitivity criteria for tunnels whose contents may be endangered by the influence of oncoming shock waves.

AN EXAMPLE

The preceding information on the motion of the tunnel walls when hit by oncoming shock wave may perhaps be better understood in the light of an example. This example is chosen to exhibit a number of considerations necessary in the prediction procedure.

It is assumed that a confined charge of $W_1 = 15$ kg is to be detonated close to a tunnel of radius $R = 1.7$ m. The arrangement is the same as that exhibited in Fig. 116 whereby $R_s = 5.0$ m. The information sought is the motion of the tunnel walls during the first impact on the tunnel of the shock wave created by the explosion.

The first pre-requisite for the solution of the problem is adequate information on the rock's properties as a transmitting medium for the stress wave. This can be obtained from a standard row experiment, and it will here be assumed that such an experiment has been performed with the following result:

Charge magnitude: 0.5 kg

Parameters: $$\kappa = 0.015, \quad \tau_0 = 1.25, \quad a = 0.4 \text{ m}$$
$$A_0^* = 15.62 \text{ m/s} \quad c_s = 5600 \text{ m/s} \qquad [14.10]$$

whereby the usual notation has been used. In addition to this the observations show a wavelength L at a distance of 5.0 m of:

$$L/c_s = 350 \times 10^{-6} \text{ s} \qquad [14.11]$$

Based on this information the first step is to predict the shock wave that will be created by the 15-kg charge. By means of eqs. 9.1, 9.3, 9.4, 9.5, 11.1, 11.2 and 11.3 the following quantities may be computed:

$$\begin{aligned}
\lambda &= -1.79282 \\
B &= 1.75776 \\
\lambda_a &= -2.04050 \\
\lambda_{a,\infty} &= -2.01474 \\
\xi_0 &= 13.78951 \\
C_\infty &= 1.74260
\end{aligned} \qquad [14.12]$$

The maximum amplitude, A, of the shock wave is now to be computed from:

$$A = A_0^* \xi^\lambda \qquad [14.13]$$

where ξ must be computed with the correct value of a for the 15-kg charge. This value is found by assuming the validity of 10.3, i.e.:

$$a = 0.4\left(\frac{15}{0.5}\right)^{1/3} = 1.2429 \text{ m} \qquad [14.14]$$

Thus, at a distance $d = R_s = 5.0$ m the correct value of $\xi = \xi_1$ would be:

$$\xi_1 = \frac{5.00}{1.2429} = 4.02 \qquad [14.15]$$

and with the computed value of λ in 14.12 and A_0^* given in 14.10 one obtains from 14.13:

$$A = 1.2895 \text{ m/s} \qquad [14.16]$$

which corresponds to A in Fig. 127.

The angle α in Fig. 127 is obtained from the information provided by the dispersion data. The meaning of the dispersion data in 14.12 is exhibited in Fig. 129. It is noticed that in the present case $\xi_1 < \xi_0$. This means that the value of $(\partial S/\partial \tau)_{max}$ computed at the distance $d = R_s$ will be obtained from:

$$\left(\frac{\partial S}{\partial \tau}\right)_{max} = C_\infty \xi_1^{\lambda_{a,\infty}} \qquad [14.17]$$

or:

$$\left(\frac{\partial S}{\partial \tau}\right)_{max} = 0.10564 \qquad [14.18]$$

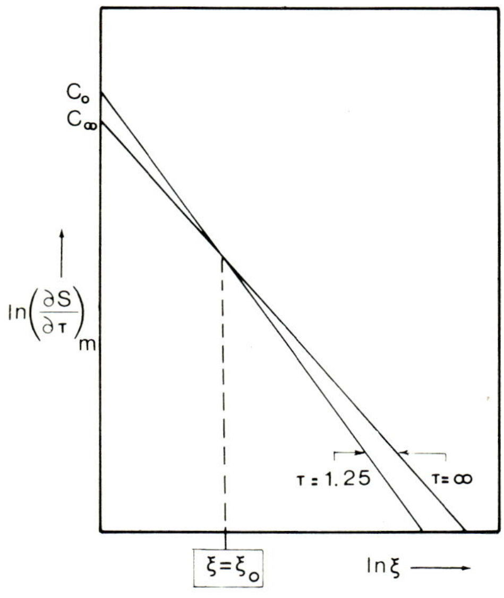

Fig. 129. Attenuation curves for the dispersion for $\tau_0 = \infty$ and $\tau_0 = 1.25$.

at $\xi = \xi_1$. The maximum amplitude of the non-dimensional stress, S, at $\xi = \xi_1$ is given by:

$$S_{max} = B\xi_1^{\lambda} = 0.14511 \quad [14.19]$$

The non-dimensional rising time, τ_r, of the oncoming shock wave at $\xi = \xi_1$ will then be given as:

$$\tau_r = \frac{S_{max}}{(\partial S/\partial \tau)_{max}} = 1.37363 \quad [14.20]$$

which in view of the definition of the non-dimensional time τ will give a corresponding rising time, t_r:

$$t_r = \frac{\tau_r a}{c_s} = 304.9 \times 10^{-6} \text{ s} \quad [14.21]$$

With t_r and A known, the value of α in Fig. 127 may easily be obtained.

The last remaining information necessary for a complete description of the oncoming shock wave is the wavelength, L, or its duration time L/c. This quantity, beside being difficult to measure because it is greatly influenced by local conditions at the pickups, is also difficult to predict by theoretical considerations. This is due to the fact that the Kelvin medium is presumably able to sustain high tension, whereas the rock is not. The theory used here thus fails to predict wavelengths with a high enough degree of accuracy, and one will have to resort to some kind of empirically obtained information.

When comparing the peak stresses σ_1 and σ_2 created at one and the same distance d by two different confined charges of magnitudes W_1 and W_2, one would by using a similarity hypothesis be able to obtain the following relation:

$$\frac{\sigma_1}{\sigma_2} = \left(\frac{W_1}{W_2}\right)^{1/3} \frac{L_1}{L_2} \quad [14.22]$$

However, if dimensional analysis is applied to the conditions at the explosion, this empirical relation may be replaced by:

$$\frac{\sigma_1}{\sigma_2} = \sqrt{\frac{a_2}{a_1}} \left(\frac{W_1}{W_2}\right)^{1/3} \frac{L_1}{L_2} \quad [14.23]$$

Here L_1 and L_2 are the wavelengths measured at a distance d from the two charges in question. Now, in view of the theory:

$$\sigma_1 = p_0 B \xi_1^{\lambda}$$
$$\sigma_2 = p_0 B \xi_2^{\lambda} \qquad [14.24]$$

where:

$$\xi_1 = \frac{d}{a_1}, \quad \xi_2 = \frac{d}{a_2} \qquad [14.25]$$

Thus:

$$\frac{\sigma_1}{\sigma_2} = \left(\frac{a_2}{a_1}\right)^{\lambda} \qquad [14.26]$$

and 14.22 and 14.23 may by assuming the validity of 10.3 be converted into:

$$\text{from 14.22:} \quad L_2 = L_1\left(\frac{a_1}{a_2}\right)^{1+\lambda}$$
$$\text{from 14.23:} \quad L_2 = L_1\left(\frac{a_1}{a_2}\right)^{\frac{1}{2}+\lambda} \qquad [14.27]$$

Using the last expression, the present case will give a duration time for the oncoming stress wave of:

$$L_2/c_s = t_0 = 350\left(\frac{0.4}{1.2429}\right)^{\frac{1}{2}-1.79282} = 1515.7 \times 10^{-6}\ \text{s} \qquad [14.28]$$

This concludes the computation of the input information necessary for the use of the formulae given in the preceding section.

The next step is to ascertain that the formulae can be used for the problem at hand, i.e. that the ratio R/L lies within the range for which the formulae have been deduced. First it is seen from 14.28 that $L = 8.488$ m with the given value of c_s, and this gives:

$$\frac{R}{L} = 0.20 \qquad [14.29]$$

which is well within the range of the validity of the formulae.

Now, the distance from the center of the explosion to the positions $\theta = 0°$ and $\theta = 90°$ of the tunnel is computed from the geometry of the problem. In view of the value of the characteristic length a of the problem [14.14], one obtains the following non-dimensional distances:

$$\xi_{0°} = 4.02\ (=\xi_1), \quad \xi_{90°} = 5.56 \qquad [14.30]$$

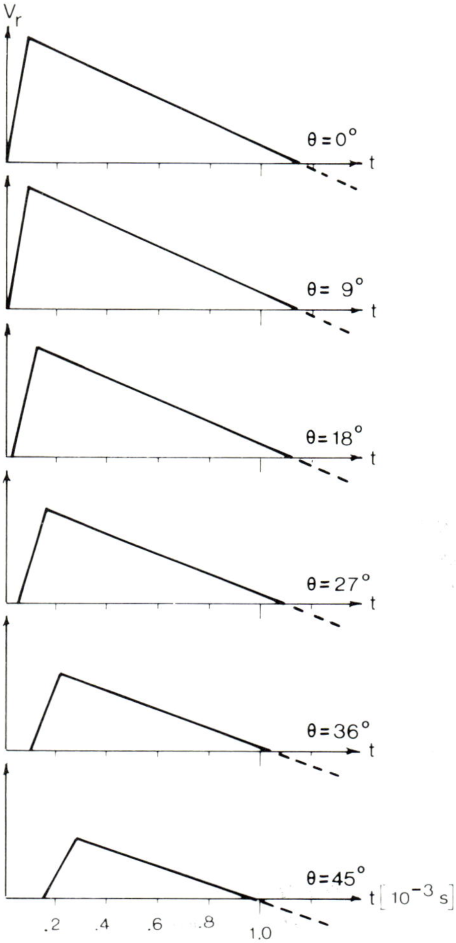

Fig. 130. Radial wall velocities computed for the example.

whereupon the ratio S_0/S_{90} of 14.5 may be computed as:

$$\frac{S_0}{S_{90}} = \frac{A_0^* \xi_{0°}^{\lambda}}{A_0^* \xi_{90°}^{\lambda}} \qquad [14.31]$$

giving the following value of the constant α in 14.5:

$$\alpha = 3.447 \qquad [14.32]$$

The evaluation of formulae 14.2 through 14.9 results in the data given in Table XXX.

TABLE XXX

Resulting data

$\theta(°)$	cta/R	$t_a(10^{-6}$ s)	V_r(m/s)	$\Delta t(10^{-6}$ s)	$\Delta t + t_a$ $(10^{-6}$ s)	t_r $(10^{-6}$ s)	$t_r + t_a$ $(10^{-6}$ s)
0	0.00000	0.0	2.579	1155.2	1155.2	68.2	68.2
9	0.01645	6.7	2.504	1140.7	1147.4	71.6	78.3
18	0.06487	26.3	2.290	1098.2	1124.5	81.4	107.7
27	0.14259	57.7	1.974	1030.8	1088.5	45.8	153.5
36	0.24566	99.4	1.603	943.4	1042.8	112.6	212.0
45	0.36929	149.5	1.227	841.9	991.4	129.1	278.6
54	0.50842	205.8	0.885	732.5	938.3	142.8	348.6
63	0.65804	266.3	0.602	621.3	887.6	151.6	417.9
72	0.81343	329.2	0.385	513.9	843.1	154.5	483.7
81	0.97032	392.7	0.233	414.4	807.1	151.2	543.9
90	1.12489	455.3	0.132	325.8	781.1	142.2	597.5

The result of the calculations is exhibited in Fig. 130 where the wall velocities in radial direction are plotted as functions of time at different locations θ around the tunnel walls. From the data one may conclude that at $\theta = 0°$, the wall will experience a maximum acceleration of:

$$a_w = 3855\,g$$

whereas the same location will have a maximum displacement inwards which is:

$$d_w = 0.15 \text{ cm}$$

This concludes the example and hopefully it exhibits the usefulness of the theoretical approach to practical problems. It may be interesting to notice that this example also has been performed as an experiment, and that the results obtained on a theoretical basis as shown here, were completely confirmed by the experimental results.

EVALUATION OF THE INVERSE LAPLACE TRANSFORM

The evaluation of the inverse Laplace transform may sometimes lead to difficulties, which, by properly deforming the path of integration, may be avoided. It is rather obvious from the way the damping parameter, κ, is introduced here (Chapter 7, section *Basic equations*) that the difficulty mainly arises from the fact that the function ζ defined in 7.13 appears as an argument in the solution for the transformed stress. In general S^* will be given as:

$$S^*(\xi, s) = f^*(s) \cdot F(A_1, \zeta) \qquad [A.1]$$

and the stress is found from:

$$S(\xi, \tau) = \frac{1}{2\pi i} \int_{b-i\infty}^{b+i\infty} f^*(s) \cdot F(A_1, \zeta)\, e^{+s\tau}\, ds \qquad [A.2]$$

where s is conceived of as the variable in the complex s-plane. The first step will be to investigate the relation between ζ and s, whereby ζ is considered to be the complex variable in the ζ-plane. Fig. A-1 shows the two planes defining s and ζ as follows:

$$\zeta = re^{i\theta}, \qquad s = \rho e^{i\varphi} - \frac{1}{\kappa} \qquad [A.3]$$

The relation between ζ and s is given as:

$$\zeta = \frac{s}{\sqrt{1 + \kappa s}} \qquad [A.4]$$

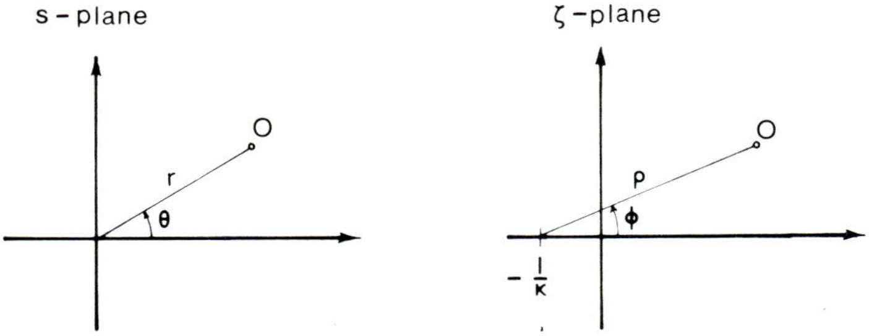

Fig. A-1. Point O in the s- and ζ-planes, respectively.

whereupon the following relations may easily be established:

$$\left.\begin{aligned} r\cos\theta &= \frac{1}{\kappa}\,[(\kappa\rho)^{1/2} - (\kappa\rho)^{-1/2}]\cos\left(\frac{\varphi}{2}\right) \\ r\sin\theta &= \frac{1}{\kappa}\,[(\kappa\rho)^{1/2} + (\kappa\rho)^{-1/2}]\sin\left(\frac{\varphi}{2}\right) \end{aligned}\right| \qquad [A.5]$$

It is easily seen that ζ as a function of s will have two Riemann sheets, and Fig. A-2 shows (by means of correspondingly shaded areas) how ζ is mapped on the two Riemann sheets in the s-plane. It should be remembered that the solution to the differential equation was established under the assumption that $Re(\zeta) > 0$. The path of integration in A.2 lies in the 1st Riemann sheet of the s-plane as shown, the only condition on it being that all singularities of the integrand lie to the left of the path.

Returning to A.1 it is noticed, that the expression for S^* appears as a product of two factors, one, $f^*(s)$, which is exclusively given by the input function and a second one, $F(A_1, \zeta)$, which is determined by the differential equation. The latter one may exhibit singularities in the left-hand half of the ζ-plane, and mapped on to the 1st Riemann sheet in the s-plane these will be located inside the circle with its center at $s = -1/\kappa$ and a radius of $1/\kappa$. The circle itself is a map of the imaginary axis in the ζ-plane between $-i2/\kappa$ and $+i2/\kappa$. The rest of the imaginary axis is

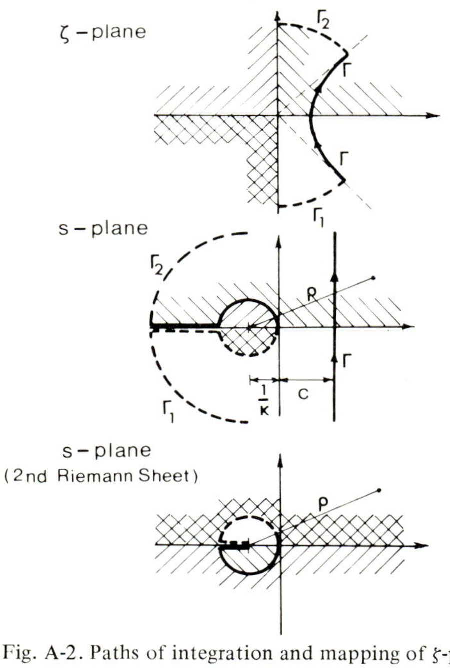

Fig. A-2. Paths of integration and mapping of ζ-plane on the s-plane.

mapped along the branch line which is located along the negative real axis in the s-plane.

With these introductory remarks on the relation between the s-plane and the ζ-plane, the evaluation of A.2 is considered. Assuming first that function $F(A_1, \zeta)$ only exhibits singularities in the left-hand half on the ζ-plane, the path of integration (Γ) is deformed on to the "big circle" (Γ_1 and Γ_2). On this path $Re(\zeta) > 0$ (as assumed in the solution) whereas $Re(s) < 0$. It is then concluded that the contribution to the integral along the big circle is negligible if the radius of it $\to \infty$. The plane has, however, a cut along the branch line (negative real s-axis) and the path must therefore go along the lower side of this cut to $s = -2/\kappa$ where it proceeds along the circle. After completing the circle, the path proceeds along the upper side of the cut back to $s = -\infty$.

Having determined the path of integration this way, the expressions for s and ζ along the path must be given:

$$\text{For } \varphi = -\pi: \quad s = -\left(\rho + \frac{1}{\kappa}\right), \quad ds = -d\rho$$
$$\zeta = \left(\rho e^{i\varphi} - \frac{1}{\kappa}\right)(\kappa\rho)^{-1/2} e^{-i\varphi/2} = -i\left(\rho + \frac{1}{\kappa}\right)(\kappa\rho)^{1/2} \qquad [A.6]$$

Introducing:

$$a = \frac{1}{\kappa}\left[(\kappa\rho)^{1/2} + (\kappa\rho)^{-1/2}\right] \qquad [A.7]$$

$$\text{one finds} \quad \zeta = -ia \qquad [A.8]$$

$$\text{For } \varphi = +\pi: \quad s = -\left(\rho + \frac{1}{\kappa}\right), \quad ds = -d\rho$$
$$\text{one finds} \quad \zeta = +ia \qquad [A.9]$$

These expressions are valid along the cut. On the circle $\kappa\rho = 1$:

$$s = \frac{1}{\kappa}(e^{i\varphi} - 1), \quad ds = \frac{i}{\kappa} e^{i\varphi} d\varphi$$
$$\zeta = \frac{2i}{\kappa} \sin\left(\frac{\varphi}{2}\right) = ib(\varphi) \qquad [A.10]$$

This shows that all along the path of integration ζ will be purely imaginary. Thus:

Along the lower part of the cut:

$$F(A_1, \zeta) = \Psi_r(a) - i\Psi_i(a) \qquad [A.11]$$

Along the circle:

$$F(A_1, \zeta) = \Psi_r(b) + i\Psi_i(b) \qquad [A.12]$$

Along the upper part of the cut:

$$F(A_1, \zeta) = \Psi_r(a) + i\Psi_i(a) \qquad [A.13]$$

The function $f^*(s)$ has so far been left out of consideration. If it is assumed that $f(t)$ is a Heaviside function:

$$f^*(s) = \frac{1}{s} \qquad [A.14]$$

This means that $f^*(s)$ has a pole at $s = 0$, i.e. on the path of integration. This difficulty is avoided through the following procedure. Let:

$$S = \frac{1}{2\pi i} \int_{c-i\infty}^{c+i\infty} \frac{1}{s} F(A_1, \zeta)\, e^{s\tau}\, ds \qquad [A.15]$$

or:

$$S^* = \frac{1}{s} F(A_1, \zeta) \qquad [A.16]$$

Then:

$$S' = \frac{\partial S}{\partial \tau} = \frac{1}{2\pi i} \int_{c-i\infty}^{c+i\infty} F(A_1, \zeta)\, e^{s\tau}\, ds \qquad [A.17]$$

where it has been assumed that differentiation under the integral sign is permissible. Then:

$$S' = -\frac{1}{2\pi i} \int_{\infty}^{1/\kappa} \{\Psi_r(a) - i\Psi_i(a)\}\, e^{-(\rho+1/\kappa)\tau}\, d\rho -$$

$$-\frac{1}{2\pi i} \int_{1/\kappa}^{\infty} \{\Psi_r(a) + i\Psi_i(a)\}\, e^{-(\rho+1/\kappa)\tau}\, d\rho +$$

$$+\frac{1}{2\pi\kappa} \int_{-\pi}^{+\pi} \{\Psi_r(b) + i\Psi_i(b)\}\, e^{-\tau[(1-\cos\varphi)-i\sin\varphi]/\kappa}\, e^{i\varphi}\, d\varphi \qquad [A.18]$$

This will after some rearrangements be:

$$S' = -\frac{1}{\pi} e^{-\tau/\kappa} \int_{1/\kappa}^{\infty} \Psi_i[a(\rho)]\, e^{-\rho\tau}\, d\rho +$$

$$+\frac{1}{\kappa\pi} \int_0^{\pi} e^{\tau(1-\cos\varphi)/\kappa} \left\{\Psi_r \cos\left(\frac{\tau}{\kappa}\sin\varphi + \varphi\right) - \Psi_i \sin\left(\frac{\tau}{\kappa}\sin\varphi + \varphi\right)\right\} d\varphi \qquad [A.19]$$

where the argument of Ψ_r and Ψ_i is b as defined as a function of φ in A.10. The final expression for S is then found by a formal integration with respect to τ of S' (see A.17):

$$S = \frac{1}{\pi} \int_{1/\kappa}^{\infty} \Psi_i[a(\rho)]\ e^{-(\rho+1/\kappa)} \frac{d\rho}{\rho + 1/\kappa} +$$

$$+ \frac{1}{2\pi} \int_0^{\pi} \frac{e^{-\tau(1-\cos\varphi)/\kappa}}{\sin(\varphi/2)} \Psi_r(b) \sin\left(\frac{\tau}{\kappa} \sin\varphi + \frac{\varphi}{2}\right) d\varphi +$$

$$+ \frac{1}{2\pi} \int_0^{\pi} \frac{e^{-\tau(1-\cos\varphi)/\kappa}}{\sin(\varphi/2)} \Psi_i(b) \cos\left(\frac{\tau}{\kappa} \sin\varphi + \frac{\varphi}{2}\right) d\varphi -$$

$$- \frac{1}{2\pi} \int_0^{\pi} \frac{\sin\varphi}{1-\cos\varphi} \Psi_i(b)\, d\varphi + \frac{1}{2\pi} \int_0^{\pi} \frac{\cos\varphi}{1-\cos\varphi} \Psi_r(b)\, d\varphi -$$

$$- \frac{1}{\pi} \int_{1/\kappa}^{\infty} \Psi_i(a) \frac{d\rho}{\rho + 1/\kappa} \qquad [A.20]$$

The last three integrals in this expression are independent of the time τ and will therefore represent the final static stage. This may be used to simplify the evaluation in cases where the final static stage is known.

This evaluation of A.2 has been carried out under the assumption that the integrand exhibits a certain analytic behavior. In cases which differ from this the procedure must be modified, but the basic reasoning behind the mapping of the ζ-plane onto the s-plane will always hold.

Appendix II

TABLES FOR S_{max}

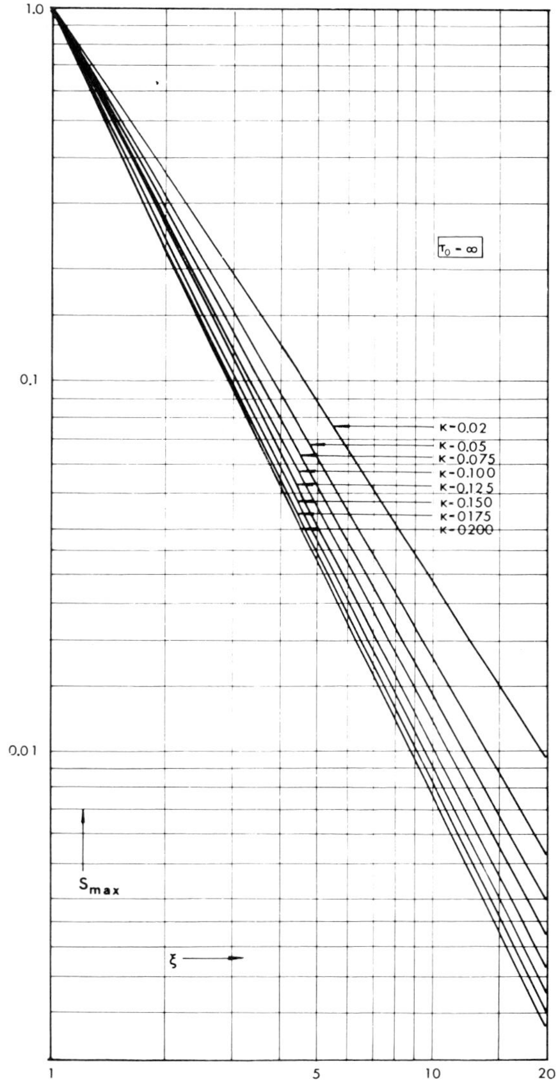

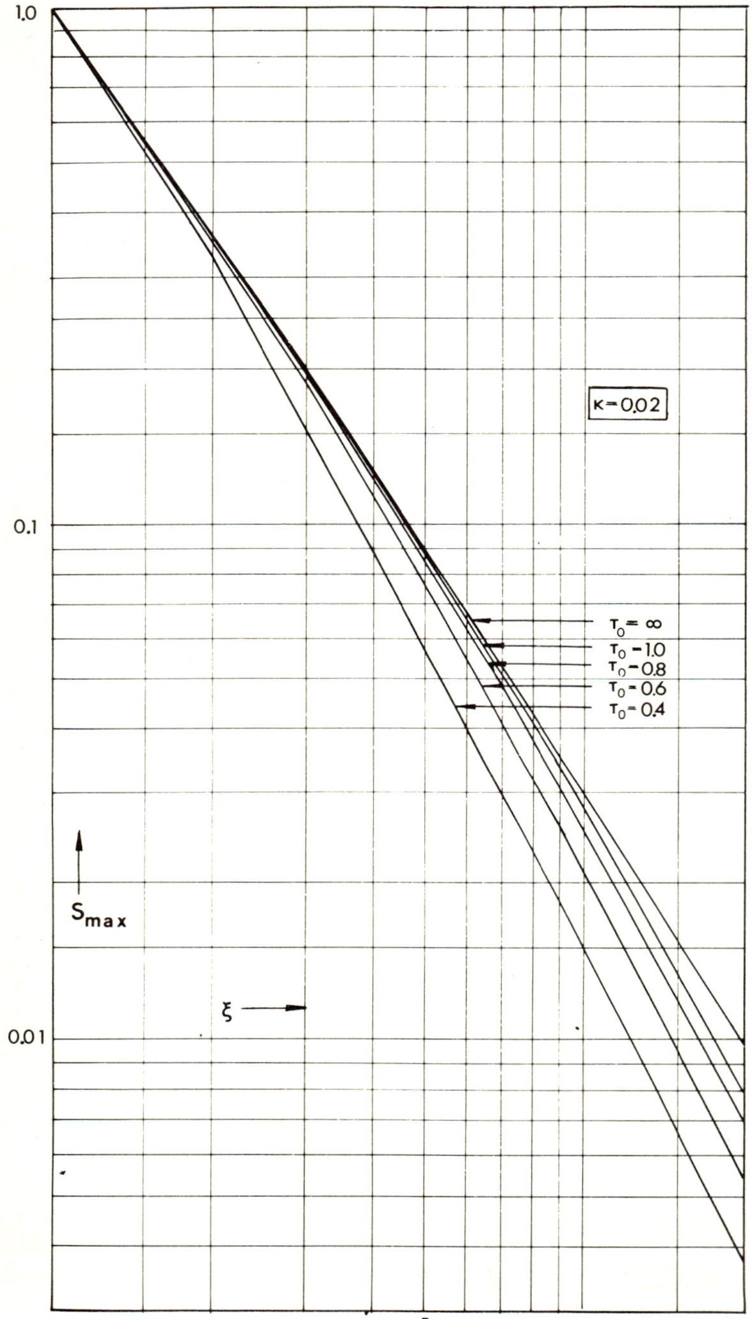
1.0
0.1
0.01
κ = 0.02
τ0 = ∞
τ0 = 1.0
τ0 = 0.8
τ0 = 0.6
τ0 = 0.4
Smax
ξ
1
5
10
20

$\kappa = 0.02$

ξ	S_{max}	τ	S_{max}	τ	S_{max}	τ
	$\tau_0 = 0.4$		$\tau_0 = 0.6$		$\tau_0 = 0.8$	
2	0.33449	1.158	0.36153	1.229	0.36574	1.255
3	0.15596	2.123	0.18793	2.193	0.19450	2.243
4	0.08898	3.089	0.11570	3.167	0.12391	3.221
5	0.05896	4.055	0.07798	4.142	0.08626	4.199
6	0.04186	5.028	0.05599	5.119	0.06358	5.178
7	0.03119	6.004	0.04144	6.093	0.04877	6.157
8	0.02408	6.982	0.03242	7.064	0.03855	7.137
9	0.01911	7.961	0.02605	8.039	0.03120	8.117
10	0.01551	8.940	0.02136	9.016	0.02544	9.097
11	0.01282	9.920	0.01781	9.995	0.02137	10.072
12	0.01076	10.900	0.01506	10.975	0.01820	11.048
13	0.00906	11.873	0.01288	11.956	0.01568	12.027
14	0.00781	12.850	0.01113	12.937	0.01363	13.007
15	0.00680	13.822	0.00971	13.919	0.01196	13.988
16	0.00597	14.801	0.00854	14.901	0.01056	14.970
17	0.00528	15.789	0.00749	15.878	0.00939	15.952
18	0.00470	16.771	0.00668	16.856	0.00841	16.933
19	0.00420	17.753	0.00599	17.835	0.00754	17.918
20	0.00378	18.736	0.00541	18.816	0.00681	18.902
	$\tau_0 = 1.0$		$\tau_0 = \infty$			
2	0.36578	1.255	0.36578	1.2547		
3	0.19531	2.252	0.19533	2.2524		
4	0.12567	3.246	0.12584	3.2497		
5	0.08870	4.235	0.08917	4.2468		
6	0.06639	5.220	0.06721	5.2428		
7	0.05169	6.202	0.05284	6.2379		
8	0.04143	7.186	0.04285	7.2323		
9	0.03394	8.168	0.03557	8.2261		
10	0.02830	9.150	0.03009	9.2195		
11	0.02394	10.133	0.02583	10.2126		
12	0.02050	11.116	0.02246	11.2054		
13	0.01759	12.097	0.01974	12.1980		
14	0.01537	13.075	0.01750	13.1905		
15	0.01355	14.055	0.01563	14.1829		
16	0.01202	15.035	0.01406	15.1752		
17	0.01074	16.016	0.01272	16.1674		
18	0.00964	16.992	0.01145			
19	0.00870	17.981	0.01058	18.1517		
20	0.00789	18.964	0.00971	19.1438		

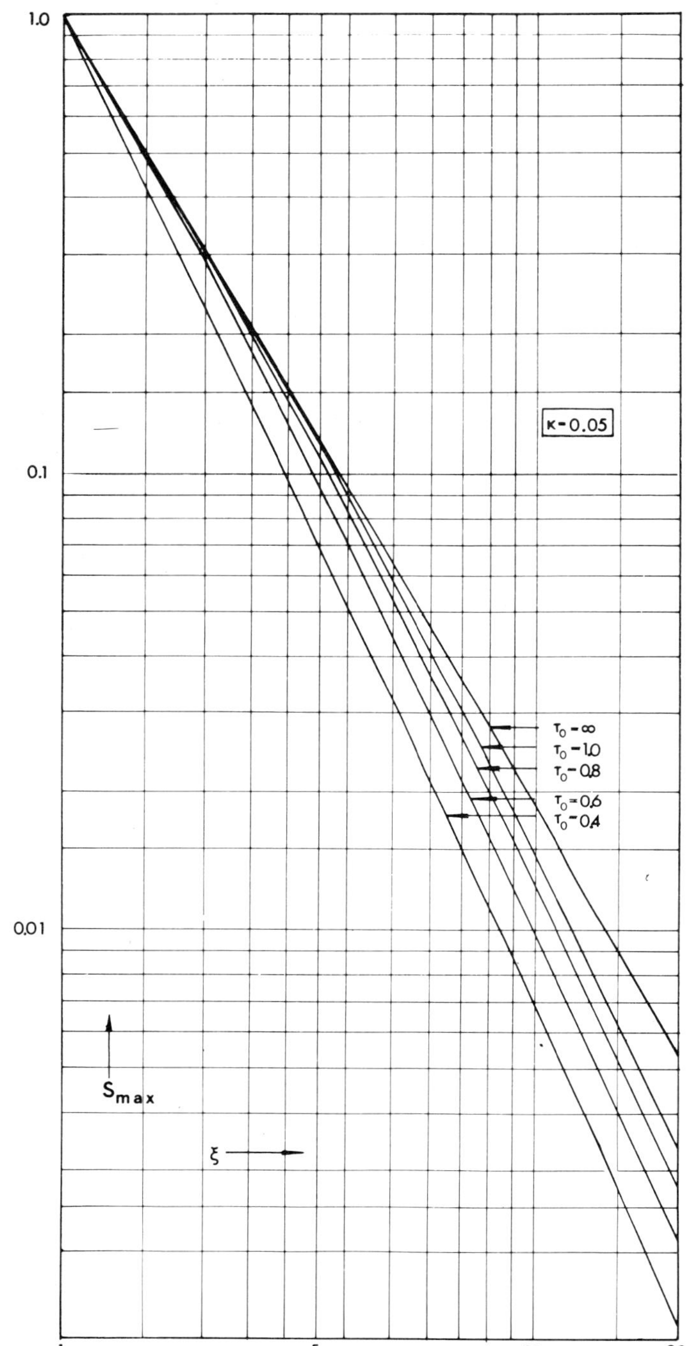
1.0
0.1
0.01
κ-0.05
τ0 - ∞
τ0 - 1.0
τ0 - 0.8
τ0 = 0.6
τ0 - 0.4
Smax
ξ
1
5
10
20

$\kappa = 0.05$

ξ	S_{max}	τ	S_{max}	τ	S_{max}	τ
	$\tau_0 = 0.4$		$\tau_0 = 0.6$		$\tau_0 = 0.8$	
2	0.23419	1.110	0.29402	1.197	0.30929	1.274
3	0.09685	2.033	0.12910	2.130	0.14663	2.202
4	0.05195	2.976	0.07053	3.062	0.08403	3.144
5	0.03195	3.9231	0.04445	4.004	0.05377	4.091
6	0.02132	4.866	0.03029	4.954	0.03699	5.031
7	0.01529	5.813	0.02181	5.907	0.02699	5.981
8	0.01144	6.766	0.01629	6.853	0.02045	6.937
9	0.00883	7.723	0.01266	7.805	0.01587	7.892
10	0.00696	8.676	0.01009	8.762	0.01271	8.842
11	0.00565	9.629	0.00820	9.722	0.01039	9.799
12	0.00466	10.587	0.00674	10.678	0.00862	10.758
13	0.00392		0.00569	11.655	0.00725	11.720
14	0.00330	12.510	0.00481	12.593	0.00615	12.679
15	0.00282	13.468	0.00413	13.555	0.00529	13.637
16	0.00244	14.426	0.00358	14.519	0.00460	14.598
17	0.00213	15.388	0.00311	15.481	0.00402	15.562
18	0.00187	16.352	0.00274	16.440	0.00354	16.527
19	0.00166	17.319	0.00243	17.403	0.00313	17.493
20	0.00147	18.283	0.00216	18.369	0.00280	18.454
	$\tau_0 = 1.0$		$\tau_0 = \infty$			
2	0.31191	1.292	0.31193	1.2921		
3	0.15288	2.258	0.15443	2.2816		
4	0.09072	3.2036	0.09398	3.2644		
5	0.05962	4.153	0.06389	4.2447		
6	0.04187	5.106	0.04655	5.2239		
7	0.03071	6.053	0.03557	6.2028		
8	0.02353	7.005	0.02814	7.1815		
9	0.01853	7.962	0.02286	8.1603		
10	0.01491	8.921	0.01897	9.1393		
11	0.01217	9.877	0.01601	10.1184		
12	0.01015	10.833	0.01369	11.0974		
13	0.00858	11.792	0.01186	12.0755		
14	0.00733	12.754	0.01037	13.0542		
15	0.00632	13.719	0.00916	14.0300		
16	0.00548	14.679	0.00814	15.0133		
17	0.00481	15.640	0.00729	15.9935		
18	0.00425	16.602	0.00657	16.9742		
19	0.00378	17.568	0.00595	17.9552		
20	0.00337	18.535	0.00541	18.9365		

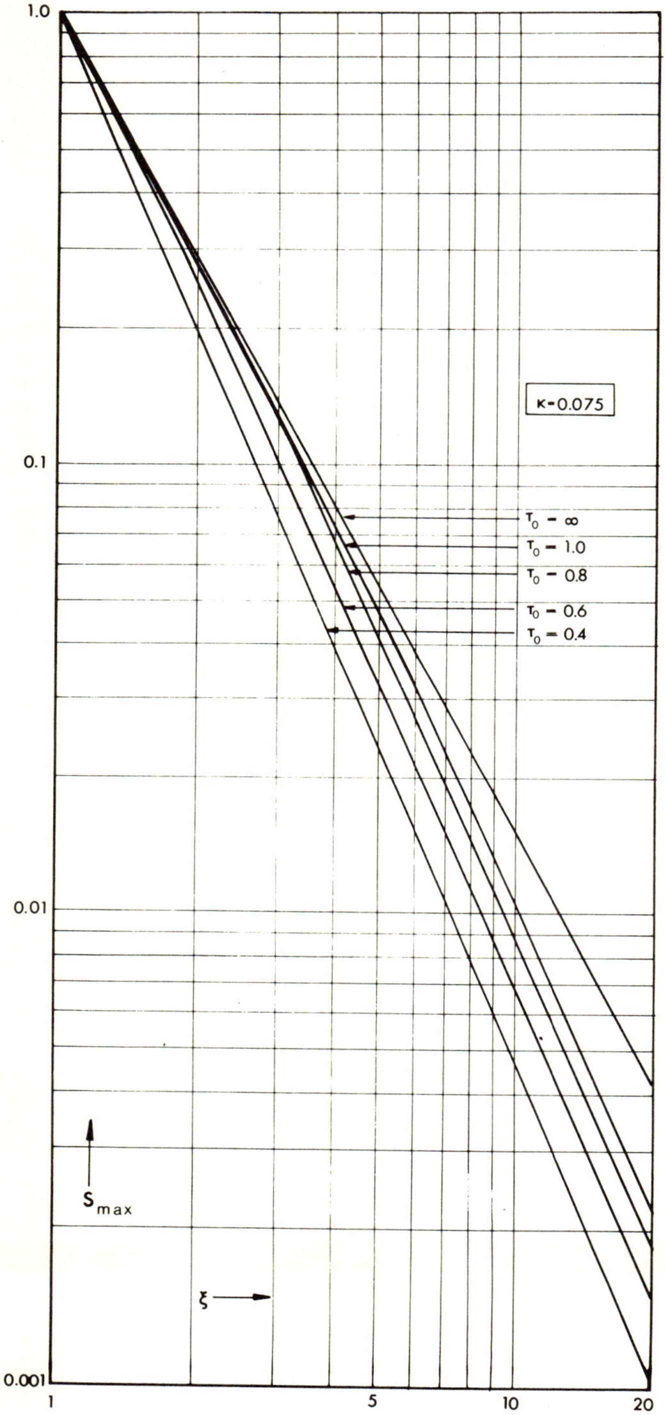
1.0
0.1
0.01
0.001
1
5
10
20
κ=0.075
τ0 = ∞
τ0 = 1.0
τ0 = 0.8
τ0 = 0.6
τ0 = 0.4
Smax
ξ

$\kappa = 0.075$

ξ	S_{max}	τ	S_{max}	τ	S_{max}	τ
	$\tau_0 = 0.4$		$\tau_0 = 0.6$		$\tau_0 = 0.8$	
2	0.19719	1.0629	0.25662	1.1688	0.27948	1.2549
3	0.07597	1.9755	0.10321	2.0656	0.12260	2.1539
4	0.03869	2.8948	0.05467	2.9821		
5	0.02331	3.8141	0.03319	3.9096	0.04107	3.9883
6	0.01538	4.7460	0.02197	4.8322	0.02760	4.9211
7	0.01074	5.6782	0.01556	5.7659	0.01960	5.8490
8	0.00792	6.6102	0.01150	6.7055	0.01460	6.7842
9	0.00605	7.5504	0.00877	7.6378	0.01124	7.7259
10	0.00472	8.4934	0.00691	8.5781	0.00884	8.6643
11	0.00379	9.4303	0.00556	9.5237	0.00714	9.6046
12	0.00302	10.3486	0.00454	10.4658	0.00586	10.5509
13	0.00258	11.3228	0.00378	11.4092	0.00489	11.5004
14	0.00216	12.2688	0.00318	12.3572	0.00411	12.4422
15	0.00184	13.2144	0.00271	13.3076	0.00351	13.3896
16	0.00158	14.1634	0.00233	14.2548	0.00303	14.3422
17	0.00137	15.1180	0.00202	15.2044	0.00263	15.2947
18	0.00120	16.0676	0.00177	16.1574	0.00230	16.2434
19	0.00105	17.0184	0.00156	17.1129	0.00203	17.1953
20	0.00093	17.9743	0.00138	18.0682	0.00180	18.1522

ξ	S_{max}	τ	S_{max}	τ
	$\tau_0 = 1.0$		$\tau_0 = \infty$	
2		1.3088	0.28308	1.3182
3			0.13607	2.2877
4			0.08052	3.2564
5	0.04667	4.0650	0.05359	4.2233
6	0.03188	4.9942	0.03847	5.1906
7	0.02295	5.9299	0.02900	6.1598
8	0.01715	6.8636	0.02260	7.1251
9	0.01328	7.8004	0.01817	8.0930
10	0.01054	8.7438	0.01494	9.0608
11	0.00851	9.6887	0.01251	10.0298
12	0.00702	10.6290	0.01063	10.9999
13	0.00587	11.5756	0.00915	11.9708
14	0.00497	12.5262	0.00796	12.9424
15	0.00424	13.4747	0.00698	13.9146
16	0.00366	14.4217	0.00618	14.8863
17	0.00319	15.3732	0.00550	15.8577
18	0.00280	16.3279	0.00494	16.8302
19	0.00247	17.2812	0.00445	17.8035
20	0.00219	18.2315	0.00404	18.7776

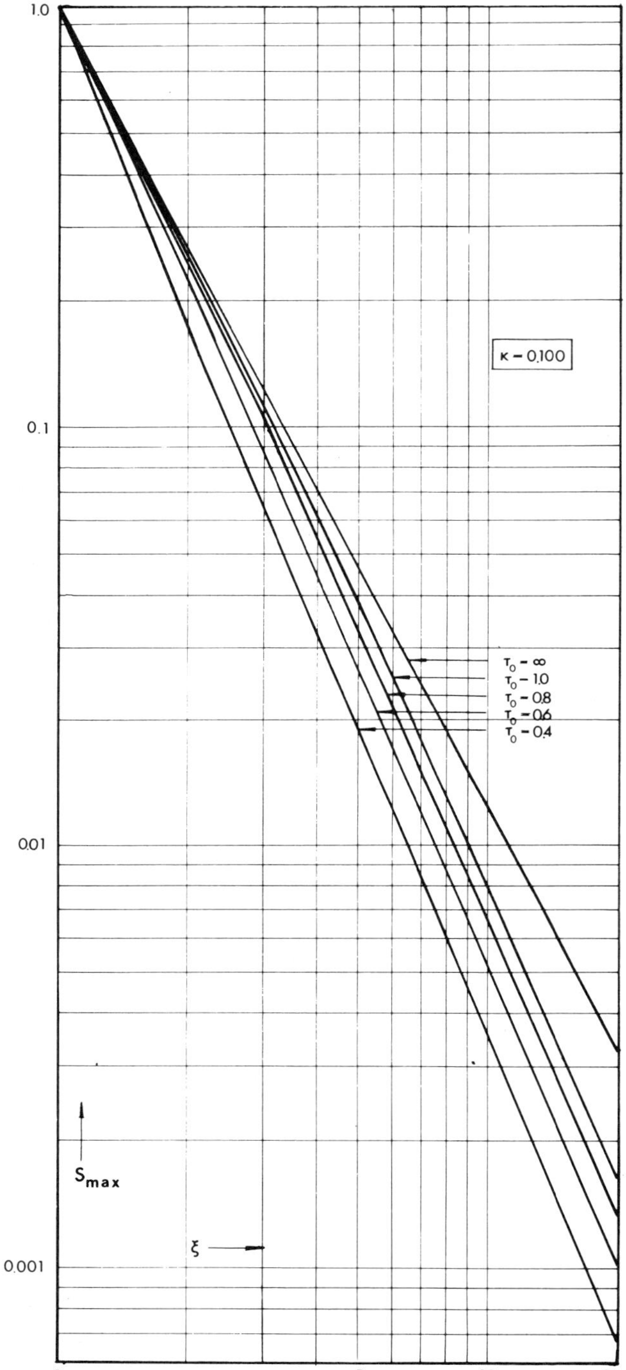
1.0
0.1
0.01
0.001
κ = 0.100
τ0 = ∞
τ0 = 1.0
τ0 = 0.8
τ0 = 0.6
τ0 = 0.4
Smax
ξ
1
5
10
20

$\kappa = 0.100$

ξ	S_{max}	τ	S_{max}	τ	S_{max}	τ
	$\tau_0 = 0.4$		$\tau_0 = 0.6$		$\tau_0 = 0.8$	
2	0.17401	1.0277	0.22859	1.1375		
3	0.06304	1.9184	0.08788	2.0093	0.10581	2.1040
4	0.03152	2.8131	0.04480	2.9103		
5	0.01853	3.7241	0.02657	3.8113	0.03340	3.9032
6	0.01198	4.6326	0.01737	4.7274	0.02199	4.8098
7	0.00832	5.5521	0.01207	5.6412	0.01544	5.7305
8	0.00604	6.4738	0.00884	6.5632	0.01131	6.6490
9	0.00457	7.3957	0.00668	7.4906	0.00862	7.5735
10	0.00356	8.3260	0.00521	8.4136	0.00674	8.5054
11	0.00283	9.2537	0.00416	9.3458	0.00539	9.4309
12	0.00230	10.1857	0.00338	10.2794	0.00439	10.3638
13	0.00190	11.1226	0.00279	11.2102	0.00364	11.3025
14	0.00159	12.0569	0.00234	12.1484	0.00305	12.2342
15	0.00134	13.0031	0.00199	13.1022	0.00260	13.1752
16	0.00115	13.9333	0.00170	14.0237	0.00222	14.1150
17	0.00099	14.8760	0.00147	14.9645	0.00192	15.0529
18	0.00086	15.8154	0.00128	15.9092	0.00167	15.9941
19	0.00076	16.7594	0.00112	16.8506	0.00147	16.9400
20	0.00067	17.7050	0.00099	17.7929		
	$\tau_0 = 1.0$		$\tau_0 = \infty$			
2			0.26580	1.3362		
3			0.12342	2.2862		
4			0.07150	3.2418		
5	0.03870	3.9790	0.04683	4.1967		
6	0.02573	4.8952	0.03313	5.1524		
7	0.01824	5.8078	0.02471	6.1091		
8	0.01350	6.7325	0.01915	7.0663		
9	0.01030	7.6558	0.01529	8.0252		
10	0.00809	8.5829	0.01249	8.9858		
11	0.00650	9.5168	0.01039	9.9477		
12	0.00530	10.4466	0.00879	10.9107		
13	0.00440	11.3810	0.00753	11.8727		
14	0.00370	12.3211	0.00652	12.8360		
15	0.00316	13.2516	0.00570	13.8009		
16	0.00270	14.1950	0.00503	14.7668		
17	0.00234	15.1409	0.00447	15.7336		
18			0.00399	16.7014		
19			0.00360	17.6614		
20			0.00325	18.6348		

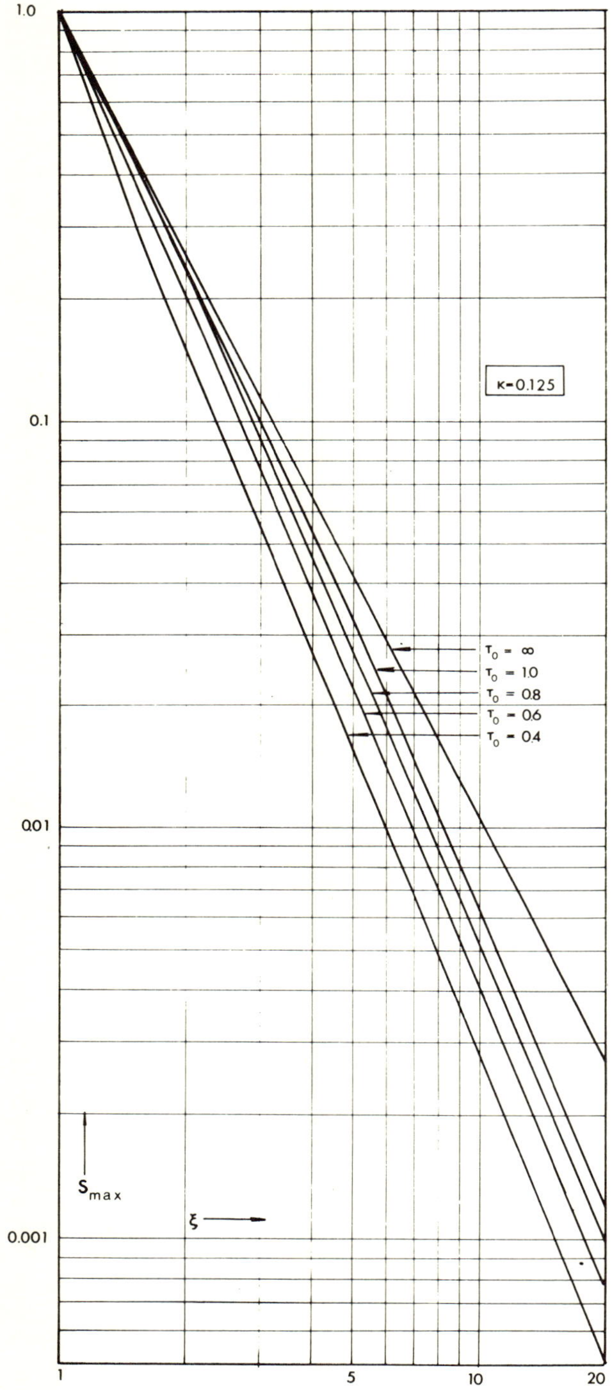
1.0
0.1
0.01
0.001
κ=0.125
τ0 = ∞
τ0 = 1.0
τ0 = 0.8
τ0 = 0.6
τ0 = 0.4
Smax
ξ
1
5
10
20

$\kappa = 0.125$

ξ	S_{max}	τ	S_{max}	τ	S_{max}	τ
	$\tau_0 = 0.4$	$\tau_0 =$	$\tau_0 = 0.6$		$\tau_0 = 0.8$	
2	0.15656	0.9959	0.20779	1.1013	0.24027	1.2005
3	0.05454	1.8583	0.07689	1.9585		
4	0.02670	2.7440				
5	0.01539	3.6323	0.02229	3.7283	0.02820	3.8136
6	0.00987	4.5336	0.01434	4.6228	0.01834	4.7150
7	0.00676	5.4347	0.00990	5.5296	0.01270	5.6145
8	0.00489	6.3457	0.00716	6.4355	0.00925	6.5264
9	0.00366	7.2577	0.00539	7.3500	0.00670	7.4362
10	0.00283	8.1729	0.00416	8.2659	0.00542	8.3537
11	0.00224	9.0950	0.00331	9.1836	0.00430	9.2744
12	0.00181	10.0126	0.00268	10.1085	0.00349	10.1934
13	0.00149	10.9384	0.00220	11.0287	0.00287	11.1204
14	0.00124	11.8658	0.00184	11.9568	0.00240	12.0452
15	0.00105	12.7921	0.00155	12.8860	0.00203	12.9734
16	0.00089	13.7253	0.00132	13.8149	0.00173	13.9058
17	0.00077	14.6550	0.00114	14.7469	0.00149	14.8368
18	0.00067	15.5889	0.00099	15.6828	0.00130	15.7706
19	0.00058	16.5267	0.00087	16.6143	0.00114	16.7069
20	0.00051	17.4636	0.00076	17.5540	0.00100	17.6421
	$\tau_0 = 1.0$		$\tau_0 = \infty$			
2			0.25223	1.3457		
3			0.11396	2.2806		
4			0.06489	3.2241		
5	0.03308	3.9019	0.04197	4.1678		
6	0.02170	4.7945	0.02940	5.1132		
7	0.01516	5.7026	0.02175	6.0605		
8	0.01108	6.6076	0.01675	7.0102		
9	0.00840	7.5229	0.01329	7.9619		
10	0.00654	8.4366	0.01081	8.9153		
11	0.00522	9.3572	0.00895	9.8687		
12	0.00423	10.2812	0.00754	10.8238		
13	0.00350	11.2031	0.00644	11.7808		
14	0.00293	12.1322	0.00556	12.7395		
15	0.00248	13.0604	0.00485	13.6992		
16	0.00212	13.9900	0.00426	14.6577		
17	0.00183	14.9236	0.00378	15.6177		
18	0.00159	15.8571	0.00337	16.5794		
19	0.00140	16.7909	0.00303	17.5423		
20	0.00123	17.7304	0.00273	18.5062		

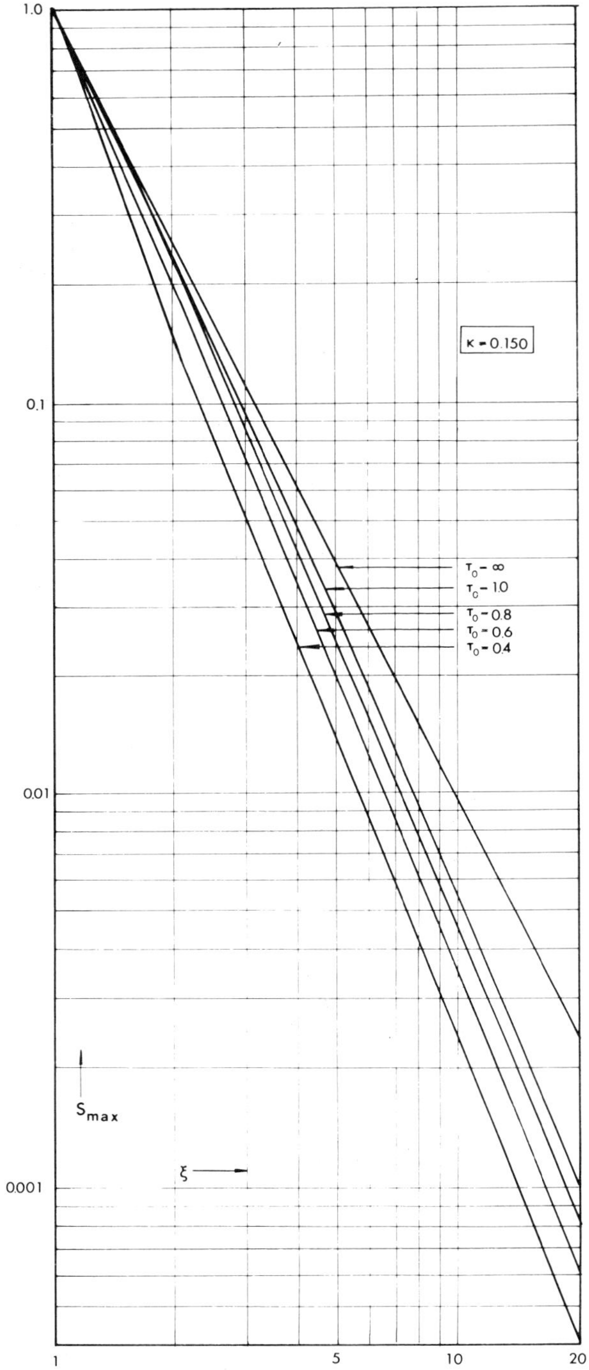
1.0
0.1
0.01
0.001
1
5
10
20
κ = 0.150
τ0 = ∞
τ0 = 1.0
τ0 = 0.8
τ0 = 0.6
τ0 = 0.4
Smax
ξ

$\kappa = 0.150$

ξ	S_{max}	τ	S_{max}	τ	S_{max}	τ
	$\tau_0 = 0.4$		*$\tau_0 = 0.6$*		*$\tau_0 = 0.8$*	
2	0.14309	0.9631	0.19369	1.0649	0.22642	1.1760
3	0.04856	1.8069	0.06864	1.9072		
4						
5	0.01325	3.5521	0.01921	3.6440	0.02452	3.7367
6	0.00838	4.4357	0.01225	4.5314	0.01572	4.6181
7	0.00571	5.3302	0.00836	5.4201	0.01081	5.5132
8	0.00409	6.2246	0.00602	6.3204	0.00779	6.4065
9	0.00305	7.1291	0.00449	7.2191	0.00584	7.3124
10	0.00235	8.0310	0.00346	8.1260	0.00451	8.2136
11	0.00185	8.9443	0.00273	9.0369	0.00357	9.1280
12	0.00149	9.8547	0.00220	9.9486	0.00288	10.0370
13	0.00122	10.7711	0.00180	10.8627	0.00236	10.9528
14	0.00101	11.6885	0.00150	11.7799	0.00197	11.8721
15	0.00085	12.6066	0.00126	12.7056	0.00166	12.7907
16	0.00072	13.5316	0.00107	13.6231	0.00141	13.7162
17	0.00062	14.4556	0.00092	14.5488	0.00121	14.6381
18	0.00054	15.3785	0.00080	15.4758	0.00105	15.5638
19	0.00047	16.3099	0.00070	16.4000	0.00092	16.4944
20	0.00041	17.2375	0.00062	17.3308	0.00081	17.4227
	$\tau_0 = 1.0$		*$\tau_0 = \infty$*			
2			0.24117	1.3499		
3			0.10663	2.2710		
4			0.05979	3.2052		
5	0.02891	3.8197	0.03826	4.1396		
6	0.01875	4.7076	0.02660	5.0742		
7	0.01296	5.5950	0.01955	6.0141		
8	0.00938	6.4960	0.01497	6.9567		
9	0.00707	7.3942	0.01182	7.9013		
10	0.00548	8.3042	0.00957	8.8465		
11	0.00434	9.2099	0.00790	9.7945		
12	0.00351	10.1254	0.00664	10.7447		
13	0.00289	11.0385	0.00565	11.6962		
14	0.00241	11.9567	0.00487	12.6474		
15	0.00203	12.8792	0.00423	13.6018		
17	0.00173	13.8000	0.00372	14.5561		
17	0.00149	14.7278	0.00329	15.5127		
18	0.00129	15.6519	0.00293	16.4688		
19	0.00113	16.5792	0.00263	17.4260		
20	0.00100	17.5143	0.00237	18.3848		

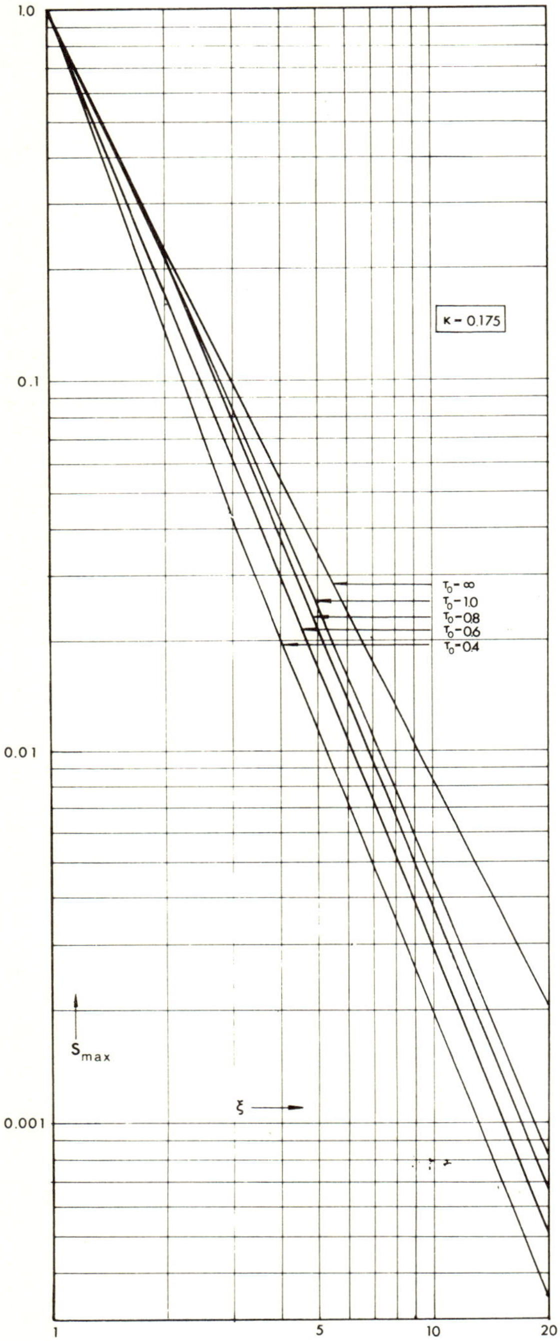
1.0
0.1
0.01
0.001
κ = 0.175
τ0 = ∞
τ0 = 1.0
τ0 = 0.8
τ0 = 0.6
τ0 = 0.4
Smax
ξ
1
5
10
20

$\kappa = 0.175$

ξ	S_{max}	τ	S_{max}	τ	S_{max}	τ
	$\tau_0 = 0.4$		$\tau_0 = 0.6$		$\tau_0 = 0.8$	
2	0.13221	0.9288	0.18339	1.0448	0.22000	1.1495
3	0.04394	1.7584				
4						
5	0.01170	3.4662	0.01698	3.5630	0.02168	3.6578
6	0.00731	4.3479	0.01070	4.4400	0.01380	4.5331
7	0.00494	5.2268	0.00726	5.3231	0.00940	5.4106
8	0.00352	6.1165	0.00518	6.2069	0.00674	6.3014
9	0.00261	7.0054	0.00385	7.1026	0.00502	7.1892
10	0.00200	7.9053	0.00295	7.9954	0.00385	8.0882
11	0.00157	8.8020	0.00232	8.8955	0.00304	8.9862
12	0.00126	9.7103	0.00186	9.8000	0.00244	9.8936
13	0.00103	10.6111	0.00152	10.7093	0.00200	10.7970
14	0.00085	11.5227	0.00126	11.6161	0.00166	11.7093
15	0.00071	12.4334	0.00106	12.5308	0.00139	12.6188
16	0.00061	13.3500	0.00090	13.4460	0.00119	13.5346
17	0.00052	14.2624	0.00077	14.3647	0.00102	14.4539
18	0.00045	15.1889	0.00067	15.2833	0.00089	15.3269
19	0.00039	16.1125	0.00058	16.2000	0.00077	16.2961
20	0.00034	17.0333	0.00051	17.1300	0.00067	17.2135
	$\tau_0 = 1.0$		$\tau_0 = \infty$			
2			0.23188	1.3538		
3			0.10046	2.2640		
4			0.05570	3.1853		
5	0.02579	3.7482	0.03533	4.1082		
6	0.01651	4.6176	0.02439	5.0371		
7	0.01133	5.5017	0.01783	5.9698		
8	0.00815	6.3844	0.01359	6.9051		
9	0.00609	7.2784	0.01069	7.8423		
10	0.00470	8.1729	0.00863	8.7827		
11	0.00371	9.0750	0.00710	9.7255		
12	0.00299	9.9776	0.00595	10.6692		
13	0.00245	10.8880	0.00505	11.6146		
14	0.00204	11.7959	0.00434	12.5624		
15	0.00171	12.7119	0.00377	13.5120		
16	0.00146	13.6245	0.00331	14.4615		
17	0.00125	14.5423	0.00292	15.4125		
18	0.00108	15.3634	0.00259	16.3654		
19	0.00095	16.3801	0.00233	17.3198		
20	0.00083	17.3067	0.00209	18.2741		

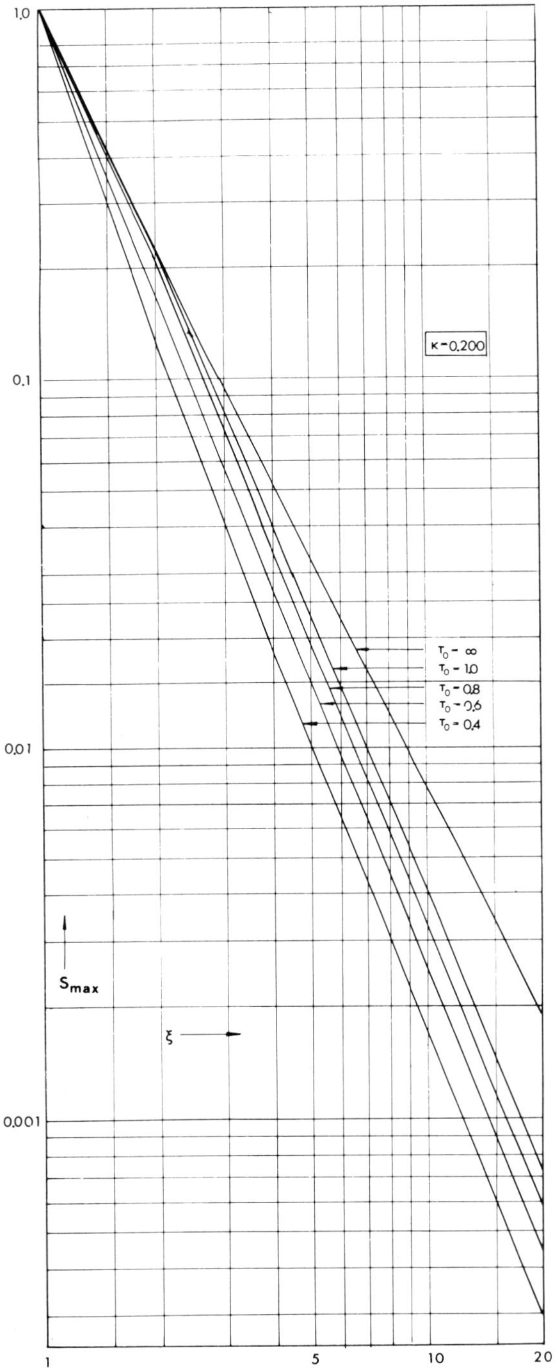
1,0
0,1
0,01
0,001
κ=0,200
τ₀ = ∞
τ₀ = 1,0
τ₀ = 0,8
τ₀ = 0,6
τ₀ = 0,4
S_max
ξ
1
5
10
20

$\kappa = 0.200$

ξ	S_{max}	τ	S_{max}	τ	S_{max}	τ
	$\tau_0 = 0.4$		$\tau_0 = 0.6$		$\tau_0 = 0.8$	
2	0.12681	0.8873	0.17376	1.0079	0.21236	1.1129
3	0.04028	1.7088	0.05770	1.8088	0.07190	1.9100
4	0.01871	2.5485	0.02710	2.6433	0.03449	2.7402
5	0.01042	3.3991	0.01520	3.4966	0.01955	3.5871
6	0.00648	4.2618	0.00952	4.3568	0.01230	4.4477
7	0.00436	5.1339	0.00641	5.2258	0.00833	5.3197
8	0.00308	6.0086	0.00455	6.1059	0.00593	6.1939
9	0.00228	6.8947	0.00337	6.9856	0.00440	7.0785
10	0.00174	7.7796	0.00257	7.8749	0.00337	7.9649
11	0.00136	8.6731	0.00202	8.7659	0.00264	8.8573
12	0.00109	9.5679	0.00161	9.6624	0.00212	9.7535
13	0.00088	10.4681	0.00131	10.5614	0.00173	10.6529
14	0.00073	11.3697	0.00109	11.4645	0.00143	11.5557
15	0.00061	12.2765	0.00091	12.3692	0.00129	12.4616
16	0.00052	13.1828	0.00077	13.2787	0.00102	13.3682
17	0.00044	14.0963	0.00066	14.1863	0.00087	14.2798
18	0.00038	15.0042	0.00057	15.1011	0.00075	15.1913
19	0.00033	15.9196	0.00050	16.0135	0.00066	16.1096
20	0.00029	16.8348	0.00044	16.9327	0.00058	17.0238
	$\tau_0 = 1.0$		$\tau_0 = \infty$			
2	0.22146	1.1549	0.22399	1.3547		
3	0.08271	1.9928	0.09541	2.2540		
4	0.04056	2.8292	0.05234	3.1650		
5	0.02328	3.6779	0.03296	4.0793		
6	0.01480	4.5379	0.02261	5.0014		
7	0.01006	5.4052	0.01644	5.9269		
8	0.00719	6.2844	0.01248	6.8552		
9	0.00536	7.1652	0.00979	7.7873		
10	0.00411	8.0531	0.00788	8.7226		
11	0.00323	8.9460	0.00647	9.6589		
12	0.00259	9.8414	0.00541	10.5979		
13	0.00212	10.7422	0.00458	11.5396		
14	0.00176	11.6442	0.00393	12.4826		
15	0.00148	12.5505	0.00341	13.4264		
16	0.00125	13.4579	0.00298	14.3727		
17	0.00107	14.3685	0.00263	15,3208		
18	0.00093	15.2838	0.00234	16.2691		
19	0.00081	16.1966	0.00209	17.2186		
20	0.00071	17.1160	0.00188	18.1698		

Appendix III

TABLES FOR $dS/d\tau|_{max}$

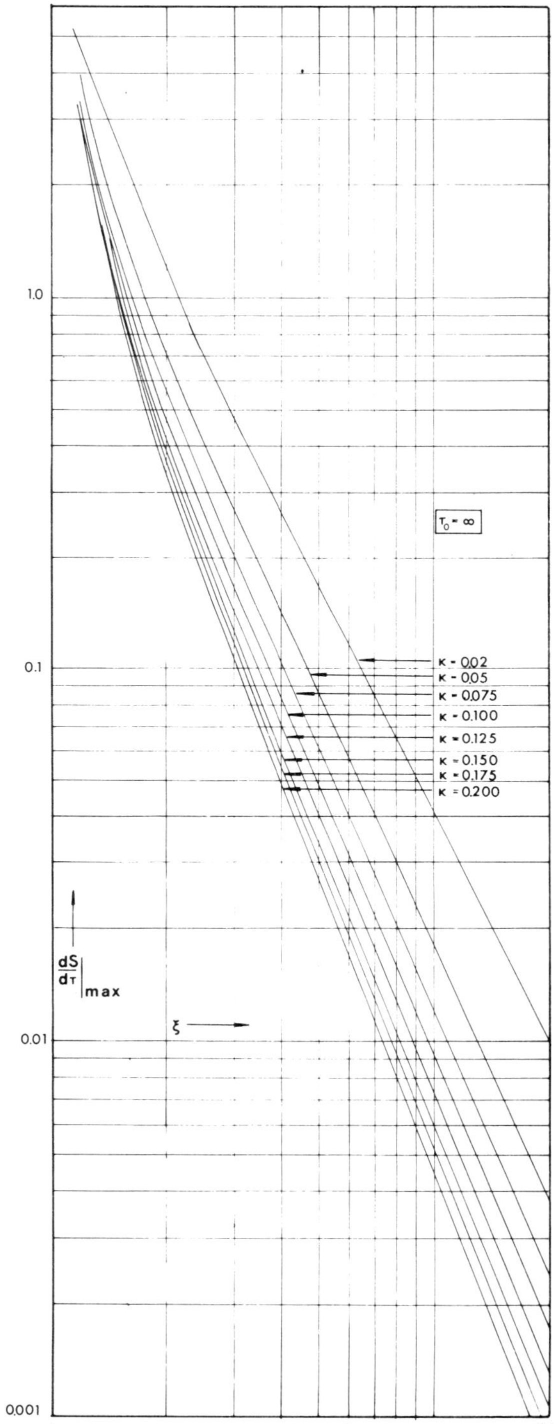

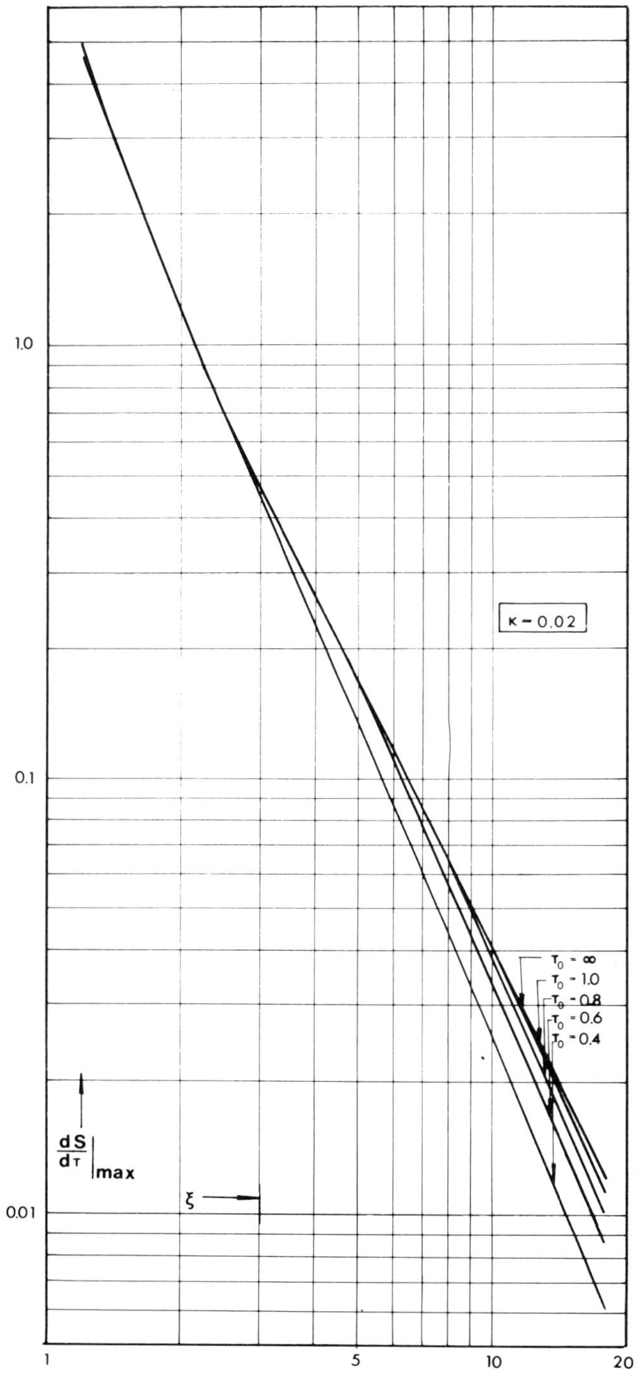
1.0
0.1
0.01
1
5
10
20
κ = 0.02
τ0 = ∞
τ0 = 1.0
τ0 = 0.8
τ0 = 0.6
τ0 = 0.4
dS/dτ |max
ξ

$\kappa = 0.02$

	$dS/d\tau|_{max}$				
ξ	$\tau_0 = 0.4$	$\tau_0 = 0.6$	$\tau_0 = 0.8$	$\tau_0 = 1.0$	$\tau_0 = \infty$
1.2	4.94880				4.64588
1.4	2.96898				2.96898
1.6	2.07347				2.07359
1.8	1.54465				1.54540
2.0	1.05933	1.06832	1.06838	1.06839	1.06839
3	0.46175	0.47694	0.47739	0.47739	0.47739
4	0.22141	0.26528	0.26693	0.26683	0.26683
5	0.12597	0.16556	0.16976	0.16973	0.16967
6	0.08518	0.11073	0.11675	0.11722	0.11710
7	0.06060	0.07439	0.08450	0.08560	0.08552
8	0.04452	0.05600	0.06342	0.06503	0.06510
9	0.03354	0.04324	0.04893	0.05085	0.05115
10	0.02589	0.03415	0.03843	0.04071	0.04121
11	0.02039	0.02738	0.03129	0.03318	0.03389
12	0.01636	0.02227	0.02582	0.02749	0.02835
13	0.01334	0.01835	0.02155	0.02316	0.02405
14	0.01116	0.01530	0.01816	0.01971	0.02066
15	0.00970	0.01314	0.01546	0.01681	0.01758
16	0.00785	0.01131	0.01335	0.01456	0.01533
17	0.00670	0.00929	0.01140	0.01255	0.01390
18	0.00584	0.00808	0.00999	0.01106	0.01212

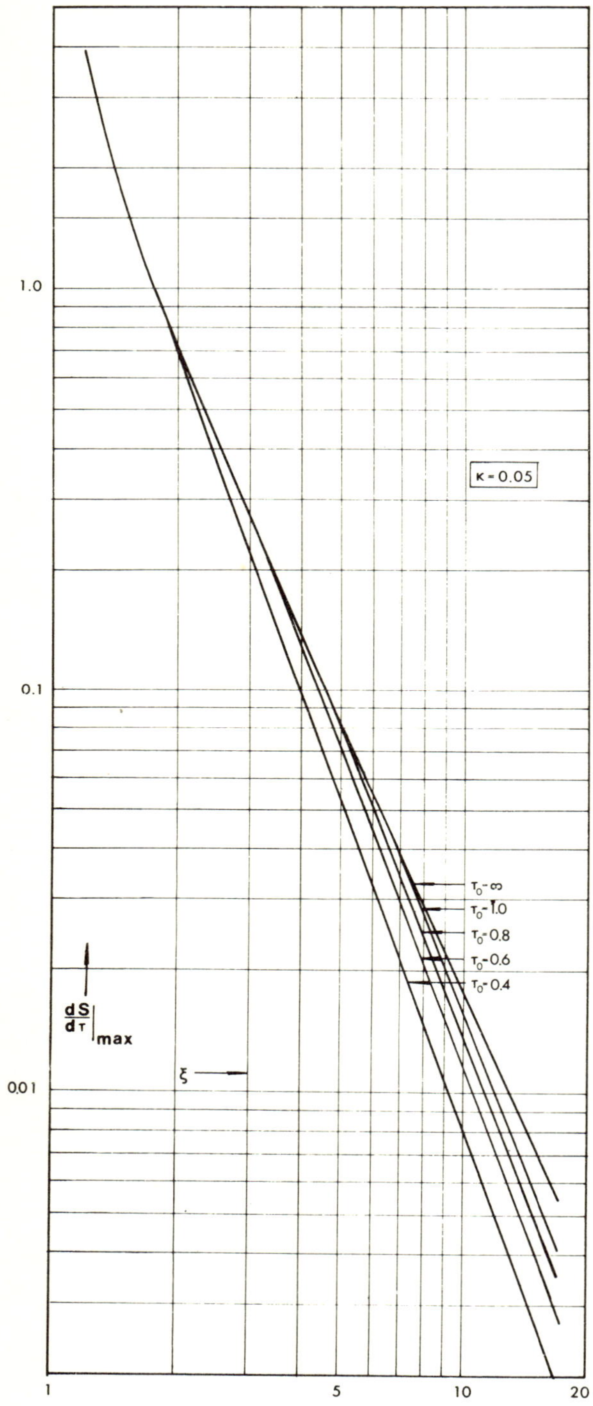
1.0
0.1
0.01
κ = 0.05
τ0 = ∞
τ0 = 1.0
τ0 = 0.8
τ0 = 0.6
τ0 = 0.4
dS/dτ|max
ξ
1
5
10
20

$\kappa = 0.050$

ξ	$dS/d\tau\|_{max}$				
	$\tau_0 = 0.4$	$\tau_0 = 0.6$	$\tau_0 = 0.8$	$\tau_0 = 1.0$	$\tau_0 = \infty$
1.2	3.72044				3.72044
1.4	1.96630				1.96630
1.6	1.29844				1.29844
1.8	0.93757				0.93757
2.0	0.69341				0.70961
5	0.05317	0.07089	0.08001	0.08371	
6	0.03209	0.05286	0.06167	0.06563	
7	0.02138		0.03489	0.03815	0.04006
8	0.01482	0.02070	0.02486	0.02755	0.02973
9	0.01072	0.01512	0.01847	0.02056	0.02283
10	0.00810	0.01141	0.01406	0.01593	0.01803
11	0.00624	0.00888	0.01095	0.01253	0.01456
12	0.00490	0.00703	0.00875	0.01003	0.01191
13	–	–	–	–	–
14	0.00323	0.00464	0.00583	0.00678	0.00845
15	0.00267	0.00385	0.00485	0.00566	0.00722
16	0.00224	0.00323	0.00409	0.00477	0.00624
17	0.00190	0.00274	0.00348	0.00406	0.00544

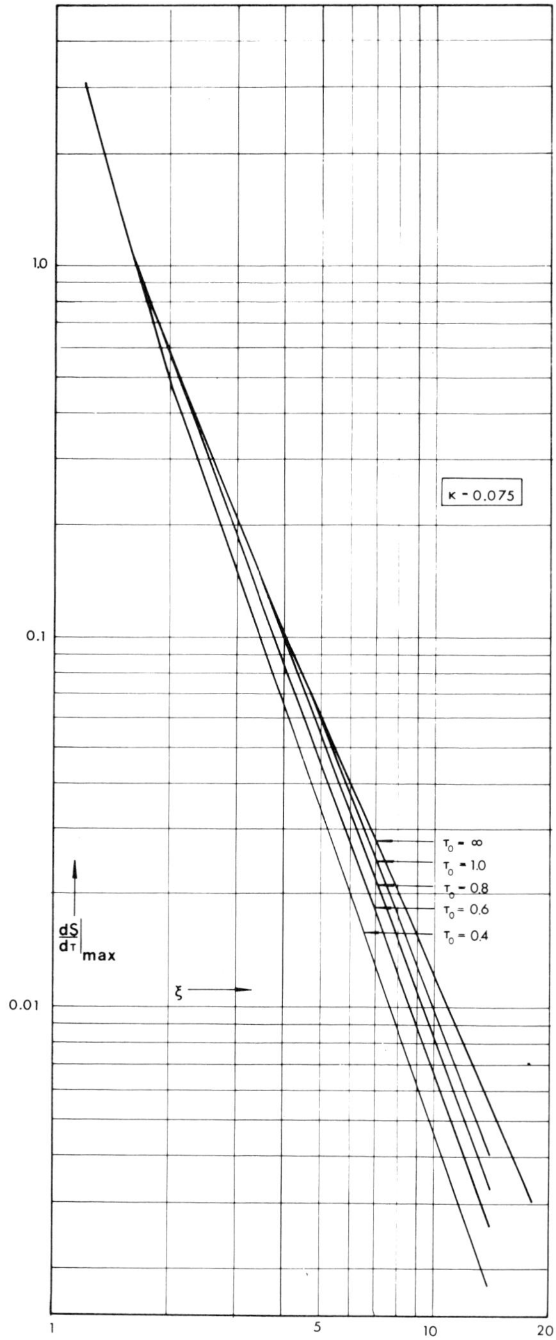
κ - 0.075
τ0 = ∞
τ0 = 1.0
τ0 = 0.8
τ0 = 0.6
τ0 = 0.4
1.0
0.1
0.01
dS/dτ|max
ξ
1
5
10
20

$\kappa = 0.075$

ξ	$dS/d\tau\|_{max}$				
	$\tau_0 = 0.4$	$\tau_0 = 0.6$	$\tau_0 = 0.8$	$\tau_0 = 1.0$	$\tau_0 = \infty$
1.2	3.16968				3.16968
1.4	1.66763				1.66763
1.6	1.06791				1.07302
1.8	0.74395				0.77783
2.0	0.47267				0.56181
5	0.03388	0.04618	0.05452	0.05799	0.06099
6	0.01996	0.02745	0.03356	0.03708	0.03983
7	0.01294	0.01816	0.02226	0.02470	0.02782
8	0.00882	0.01257	0.01546	0.01768	0.02044
9	0.00633	0.00902	0.01129	0.01299	0.01551
10	0.00469	0.00674	0.00844	0.00982	0.01212
11	0.00358	0.00516	0.00652	0.00758	0.00969
12	0.00280	0.00404	0.00512	0.00598	0.00792
13	0.00222	0.00323	0.00409	0.00481	0.00656
14	0.00180	0.00262	0.00332	0.00395	0.00550

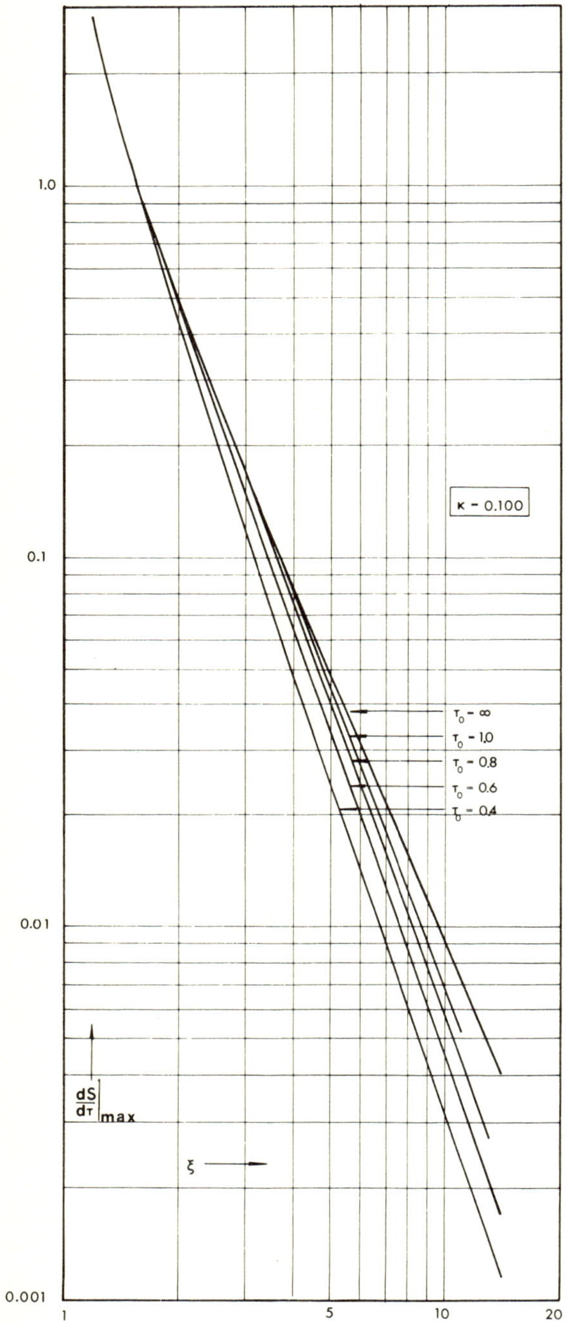

1.0
0.1
0.01
0.001
1
5
10
20
κ - 0.100
τ0 - ∞
τ0 - 1.0
τ0 - 0.8
τ0 - 0.6
τ0 - 0.4
dS/dτ |max
ξ

$\kappa = 0.100$

ξ	$dS/d\tau\|_{max}$				
	$\tau_0 = 0.4$	$\tau_0 = 0.6$	$\tau_0 = 0.8$	$\tau_0 = 1.0$	$\tau_0 = \infty$
1.2	2.81294				2.81294
1.4	1.52088				1.52088
1.6	0.94125				0.94153
1.8	0.64086				0.65449
2.0	0.42312				0.47381
5	0.02440	0.03359	0.04092	0.04498	0.04790
6	0.01407	0.02005	0.02450	0.02773	0.03097
7	0.00900	0.01277	0.01596	0.01819	0.02135
8	0.00606	0.00864	0.01093	0.01267	0.01548
9	0.00431	0.00620	0.00785	0.00912	0.01168
10	0.00315	0.00458	0.00580	0.00682	0.00906
11	0.00239	0.00347	0.00440	0.00529	0.00720
12	0.00185	0.00268	0.00344	–	0.00583
13	0.00146	0.00212	0.00279	–	0.00480
14	0.00117	0.00172	–	–	0.00402

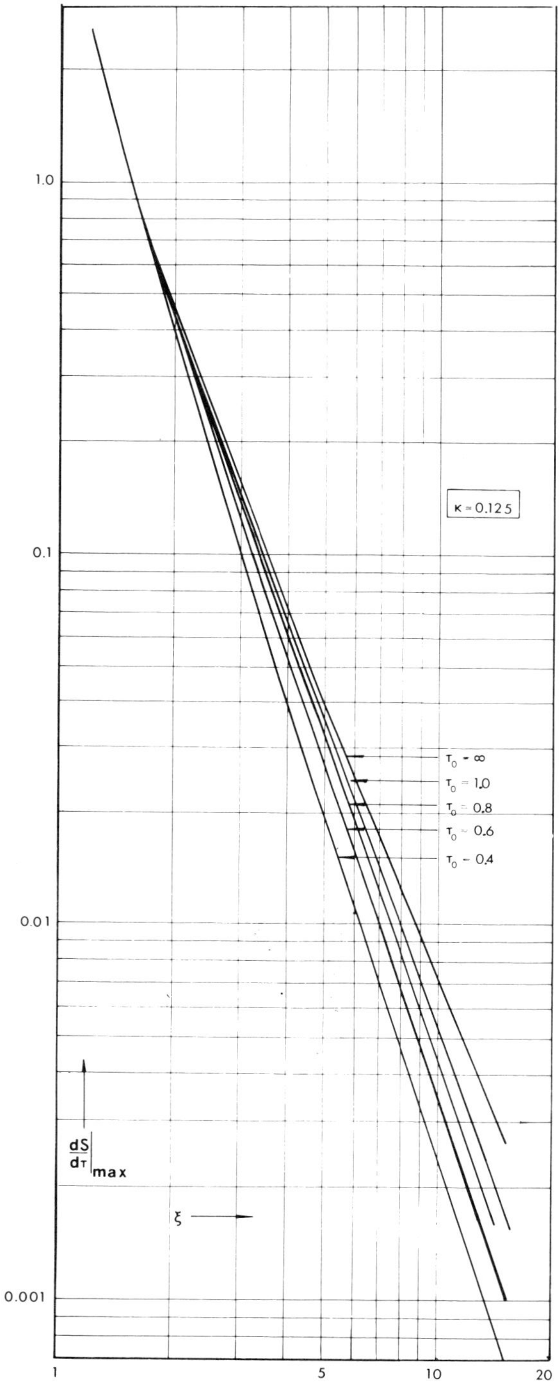
1.0
0.1
0.01
0.001
1
5
10
20
κ = 0.125
τ0 = ∞
τ0 = 1.0
τ0 = 0.8
τ0 = 0.6
τ0 = 0.4
dS/dτ|max
ξ

$\kappa = 0.125$

ξ	$dS/d\tau\|_{max}$				
	$\tau_0 = 0.4$	$\tau_0 = 0.6$	$\tau_0 = 0.8$	$\tau_0 = 1.0$	$\tau_0 = \infty$
1.2	2.64592				2.64592
1.4	1.40911				1.40946
1.6	0.85636				0.85763
1.8	0.57048				0.58838
2.0	0.38738	0.42111	0.42691	0.42693	0.42693
5	0.01892	0.02666	0.03240	0.03640	0.03986
6	0.01085	0.01542	0.01915	0.02191	0.02529
7	0.00679	0.00975	0.01228	0.01416	0.01730
8	0.00453	0.00656	0.00828	0.00972	0.01246
9	0.00318	0.00461	0.00589	0.00694	0.00931
10	0.00232	0.00338	0.00432	0.00513	0.00719
11	0.00174	0.00254	0.00327	0.00390	0.00568
12	0.00134	0.00197	0.00253	0.00303	0.00459
13	0.00106	0.00155	0.00200	0.00241	0.00377
14	0.00085	0.00124	0.00161	0.00194	0.00313
15	0.00069	0.00101	0.00131	0.00159	0.00264

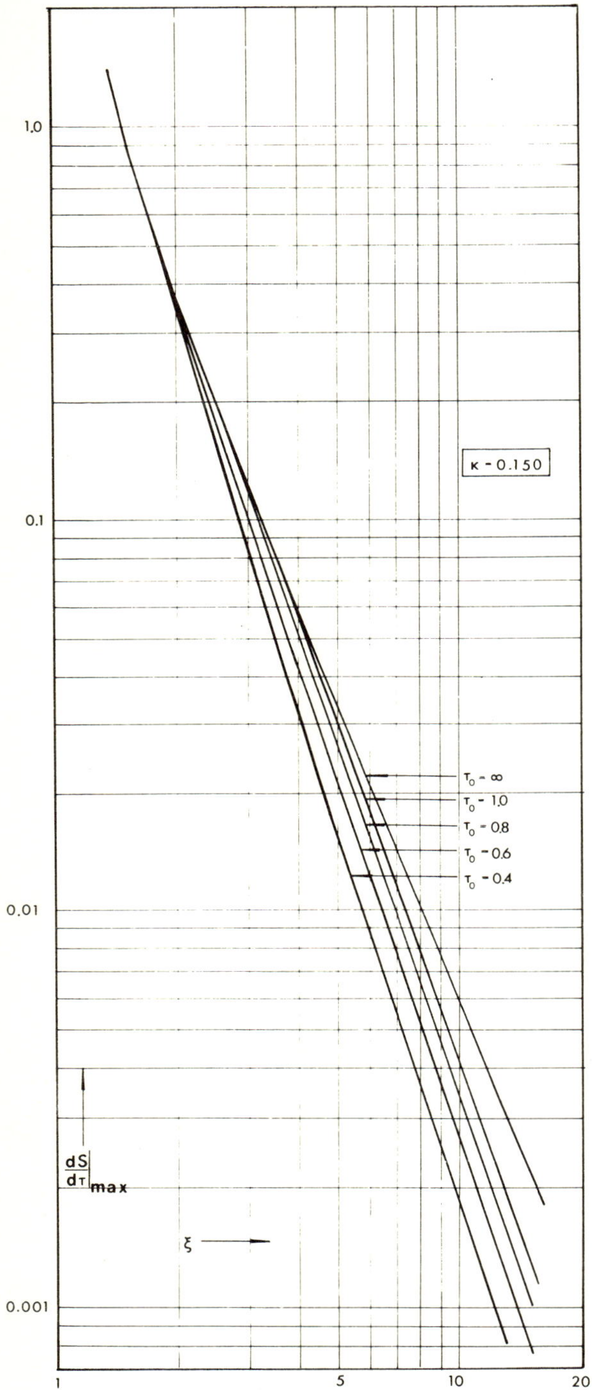
1.0
0.1
0.01
0.001
κ - 0.150
τ0 - ∞
τ0 - 1.0
τ0 - 0.8
τ0 - 0.6
τ0 - 0.4
dS/dτ|max
ξ
1
5
10
20

$\kappa = 0.150$

ξ	$dS/d\tau\|_{max}$				
	$\tau_0 = 0.4$	$\tau_0 = 0.6$	$\tau_0 = 0.8$	$\tau_0 = 1.0$	$\tau_0 = \infty$
1.2	2.85945				2.85945
1.4	1.37815				1.37815
1.6	0.81921				0.81911
1.8	0.54126				0.54126
2.0	0.35299				0.38328
5	0.01545	0.02179	0.02692	0.03040	0.03398
6	0.00871	0.01247	0.01561	0.01797	0.02145
7	0.00539	0.00779	0.00982	0.01150	0.01455
8	0.00357	0.00517	0.00660	0.00780	0.01040
9	0.00249	0.00363	0.00465	0.00551	0.00774
10	0.00179	0.00263	0.00338	0.00404	0.00596
11	0.00134	0.00197	0.00255	0.00305	0.00468
12	0.00103	0.00152	0.00196	0.00236	0.00376
13	0.00081	0.00119	0.00154	0.00186	0.00308
14	–	0.00095	0.00123	0.00149	0.00255
15	–	0.00077	0.00100	0.00121	0.00214

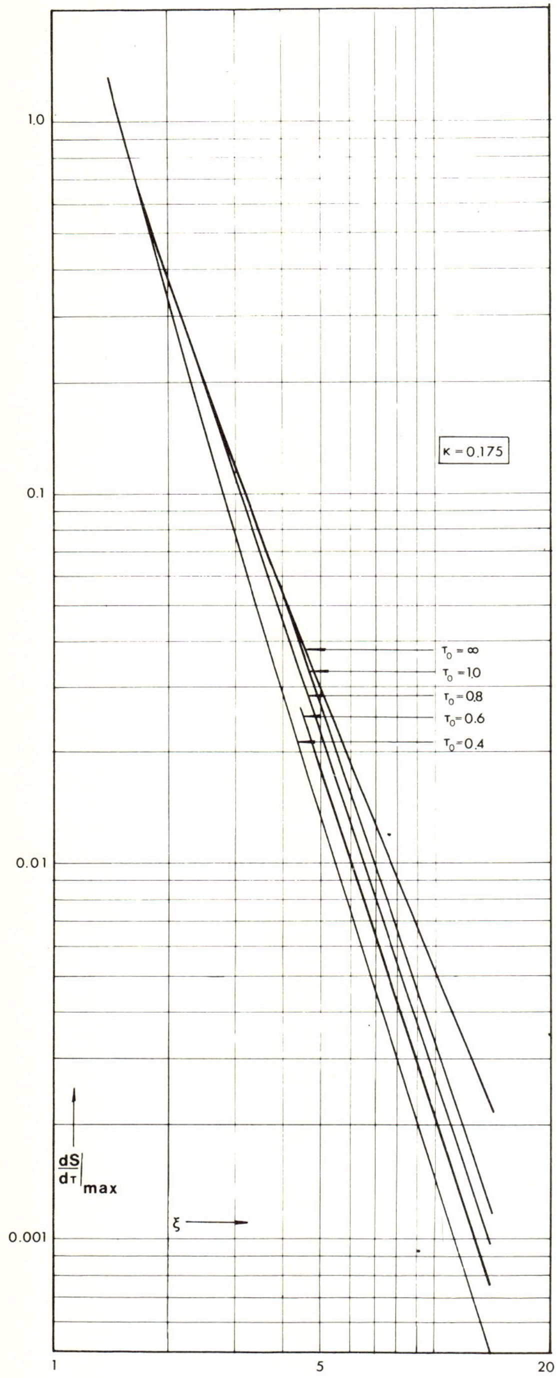

1.0
0.1
0.01
0.001
κ = 0.175
τ0 = ∞
τ0 = 1.0
τ0 = 0.8
τ0 = 0.6
τ0 = 0.4
dS/dτ|max
ξ
1
5
20

$\kappa = 0.175$

ξ	$dS/d\tau\|_{max}$				
	$\tau_0 = 0.4$	$\tau_0 = 0.6$	$\tau_0 = 0.8$	$\tau_0 = 1.0$	$\tau_0 = \infty$
1.2					
1.4	1.28711				1.34680
1.6	0.76383				0.76466
1.8	0.50400				0.50580
2.0	0.32160				0.36456
5	0.01298	0.01846	0.02289	0.02650	0.03031
6	0.00724	0.01043	0.01310	0.01525	0.01861
7	0.00444	0.00644	0.00818	0.00960	0.01256
8	0.00291	0.00424	0.00544	0.00644	0.00893
9	0.00202	0.00295	0.00380	0.00453	0.00661
10	0.00146	0.00213	0.00275	0.00330	0.00505
11	0.00108	0.00159	0.00206	0.00248	0.00396
12	0.00082	0.00122	0.00158	0.00191	0.00317
13	–	0.00095	0.00123	0.00149	0.00259
14	0.00050	0.00075	0.00098	0.00119	0.00214

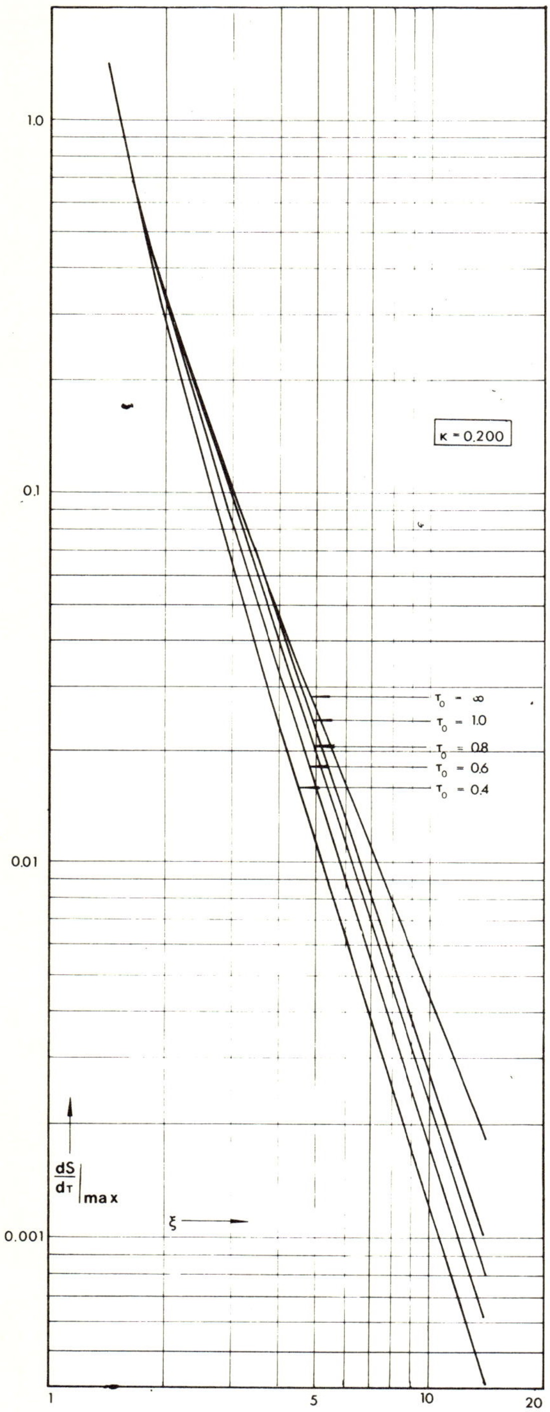
1.0
0.1
0.01
0.001
κ = 0.200
τ0 = ∞
τ0 = 1.0
τ0 = 0.8
τ0 = 0.6
τ0 = 0.4
dS/dτ |max
ξ
1
5
10
20

$\kappa = 0.200$

ξ	$dS/d\tau\|_{max}$				
	$\tau_0 = 0.4$	$\tau_0 = 0.6$	$\tau_0 = 0.8$	$\tau_0 = 1.0$	$\tau_0 = \infty$
1.2					
1.4	1.40460				1.40460
1.6	0.74511				0.74511
1.8	0.48200				0.48364
2.0	0.29503				0.33997
3	0.06548	0.08729	0.10073	0.10463	0.10491
4	0.02382	0.03331	0.04055	0.04508	0.04800
5	0.01128	0.01607	0.02004	0.02298	0.02661
6	0.00620	0.00893	0.01131	0.01318	0.01651
7	0.00377	0.00547	0.00698	0.00824	0.01106
8	0.00245	0.00359	0.00460	0.00548	0.00782
9	0.00169	0.00247	0.00319	0.00383	0.00576
10	0.00121	0.00177	0.00230	0.00277	0.00439
11	0.00089	0.00131	0.00170	0.00206	0.00343
12	0.00068	0.00100	0.00130	0.00157	0.00274
13	0.00052	0.00078	0.00102	0.00124	0.00223
14	0.00041	0.00062	0.00082	0.00101	0.00184

REFERENCES

Baron, M.L. and Matthews, A.T., 1961. Diffraction of a pressure wave by a cylindrical cavity in an elastic medium. *J. Appl. Mech.*, 28: 347–354.

Baron, M.L. and Parnes, R., 1962. Displacements and velocities produced by the diffraction of a pressure wave by a cylindrical cavity in an elastic medium. *J. Appl. Mech.*, 29: 385–394.

Batemann Manuscript Project, 1954. *Tables of Integral Transforms, 1.* McGraw-Hill, New York, N.Y.

Batemann Manuscript Project, 1953. *Higher Transcendental Functions, 2.* McGraw-Hill, New York, N.Y.

Kestin, J., 1968. *A Course in Thermodynamics, 2.* Ginn Blaisdeli, Waltham, Mass.-Toronto-London.

Persen, L.N., 1968. *Über die grundlegenden felsdynamischen Beziehungen die bei Schutzbauwerken im Felsen zu beachten sind.* Acta Polytechnica Scandinavica, Stockholm.

Selberg, H.I., 1952. Transient compression waves from spherical and cylindrical cavities. *Ark. Fys.*, 5(7): 97–108.

INDEX